The Goldfish and Its Systematic Culture
with a View to Profit

The Japanese fantail goldfish (*Carassius Japonicus*)

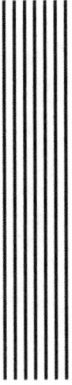

# The Goldfish and Its Systematic Culture with a View to Profit

*A Practical Treatise on the Fish, Its Propagation, Enemies, Diseases, and Care of the Fish in Captivity, Together with Hints on the Construction of Ponds, Etc.*

*HUGO MULERTT*
*Propagator of Ornamental Fish and Aquatic Plants*

WAKING LION PRESS

ISBN 978-1-4341-0386-4

This book was first published in 1883. This edition has been completely redesigned and reset for modern readers.

Published by Waking Lion Press, an imprint of The Editorium

Waking Lion Press™, the Waking Lion Press logo, and The Editorium™ are trademarks of The Editorium, LLC

The Editorium, LLC
West Jordan, UT 84081-6132
wakinglionpress.com
wakinglion@editorium.com

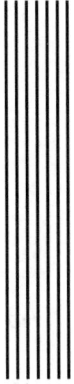

# Contents

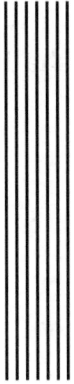

# Preface

There are not many, if any, homes in the country or suburbs of large cities where fish culture could not be indulged in, either as an agreeable pastime or with a view of realizing an income from it.

Being accidently drawn into the culture of ornamental fish through a genuine love of Natural History, I have, for the past ten years, given it my exclusive attention.

The originality of the method I describe in the following pages I claim as my own, and it has been my earnest endeavor to speak plainly and to the point, giving the full results of my long experience, not intentionally withholding any facts or secrets that would in any way have a bearing upon the success of the business.

This method has the decided advantage of applying equally to the culture of goldfish, no matter upon what scale it is carried on, whether the establishment is intended to produce hundreds of thousands of fish annually, or be engaged in with limited means by the deserving widow or invalid who wishes to add something to a restricted income.

With these objects in view, and to promote a more thorough knowledge of the goldfish, as well as to stimulate the study of Natural History, I have made every exertion, and should my efforts meet with the approval of an intelligent public, I shall consider that I have not labored in vain.

HUGO MULERTT.

CINCINNATI, Ohio, July, 1883.

*Part I*

# THE GOLDFISH AND ITS VARIETIES

# 1 The History of the Goldfish

*"Every man is a valuable member of society, who, by his observations, researches, and experiments, procures knowledge for men."*
*—Smithson.*

Long before the principles upon which the success of an aquarium are based were understood, goldfish were kept and tenderly cared for, merely because they are beautiful, and besides being in a sense pets, upon. which to lavish much solicitude and attention, served the purpose of an animated ornament, than which it is difficult to find a more beautiful and desirable substitute. The very fact of their being denizens of a different element than that in which the most familiar domestic animals and pets live, served to surround them with a halo of fascination that prompted their admirers to expend upon them any amount of time and pains to keep them in a flourishing condition, and ever ready to be displayed before the chosen guest in the household.

Their wonderful tenacity of life, the gorgeous colors in which they are clothed, both contributed to make them adapted to life in the aquarium and grace the elaborate structures that were framed for the express purpose of displaying them to the best advantage.

The goldfish, we are told, like many other things, originated in China, though, so far as this is concerned, its history is involved in much obscurity. The first mention we find of it is in the "Arabian Nights' Entertainments," in the story of the Fisherman and the Genii, a story, which like those accompanying it, is of Persian origin. It is, however, frequently represented in the old Chinese paintings, and appears to have been held in great esteem, insomuch as several of the ancient and distinguished families among the Chinese carried it as a component part of the family coat of arms, or as an ornament upon their armor. In short, the goldfish has always been looked upon as a kind of superior being among the finny tribe, the ancients even dedicating it to their well-beloved goddess, Venus.

In China, even at the present day, the goldfish enjoys the admiration of all, and may be seen in almost every house, inhabiting a richly-decorated China bowl, or disporting itself in lakelets in the gardens, made for its especial care and enjoyment. They are alike admired and beloved from the august Emperor on his throne of State down to the most humble and impoverished peasant in the realm, all seeing in it an object worthy of care, and, it might be said, of love. So intense is the fondness for the goldfish in the land of Chinamen that it has begotten a study of them and their habits, that amounts to a science. All the resources of the knowledge of them, thus gained, have been exhausted in the production of new varieties by the crossing of different species, and with marked success.

In the larger cities the fish are regularly brought to market and offered for sale, very much as our florists offer their flowers and plants. For that purpose they are separated into pairs, and placed in little vessels made of bamboo, and, together with some bit of water plant, are sold to ready and ever eager purchasers; the prices ranging according to the variety and perfection of the specimens.

The mandarin, in person, may sometimes be seen in the markets buying fish for his aquaria, which, indeed, are often made of carved ivory, and inlaid with gold. These purchases he makes with the same relish and delight that one takes in making a new and valuable addition to his conservatory.

The most choice and beautiful varieties are obtained from a lake in the province of Che-Kyang.

The first introduction of the goldfish into Europe is variously dated, the years 1691 and 1728, A.D., each having claims for that honor; it may be said also that the variety introduced was the poorest and cheapest the Chinese had.

The first seen in France, however, were those imported for the famous Madame de Pompadour. Soon afterwards they became quite common, as it was found that they throve well in the waters of Southern Europe, especially in Portugal, where they sprang from a few small fish, said to have escaped from a vessel newly arrived from China. In that country they are now considered a delicacy for the table. It was not long until several streams in the neighborhood of Lisbon absolutely swarmed with them, and it is from this source that all Europe became stocked with them. From Europe the fish were brought to America, and quickly won their way into favor. For the last forty years, perhaps, these fish have become wild in the United States. Some having accidentally escaped into open waters, they soon made themselves at home, became thoroughly acclimated, and in consequence the goldfish has been quoted by several authors as a native American fish. It is true that, having been born here, they are in one sense native, but are *not native* in the sense, origin, as having originated here.

Living thus in a wild state, the fish has greatly degenerated from the original standard. Through the efforts of Admiral ———, U. S. N., the cultivation of the goldfish in the United States has received a new impulse.

This gentleman, but a few years ago, brought from Japan a number of specimens of the choicest varieties which have since been reproduced with marked success.

The annual sale of goldfish in this country at the present day may be estimated at two millions, and of a value of $300,000.

Although this number is greatly exceeded by Europe, ours are their superior in quality, and the *wholesale* prices received for them are from one hundred to five hundred per centum higher than those paid to European culturists, who are at present endeavoring to compete in the American market.

# 2 The Common Goldfish

*(Carassius auratus.) Der goldfisch. Le poisson rouge.*

The goldfish belongs to the carp-family, and by some authors is called *Cyprinus auratus,* which means gold-carp. A thorough investigation of the subject has led the author to the adoption of the name used in the heading, which is also used by some other authors. Besides being more correct, according to the anatomical structure of the fish, is more distinguishing, as there exists a variety of the European food-carp, known as the golden carp, or *cyprinus aureus,* which is in no ways identical with the goldfish *carassius auratus.*

The body of the fish is elongated, compressed upon the sides and entirely covered with a coat of uniformly-sized scales. The head is short, naked, that is, without scales; the lips well developed and without barbles. The color is generally an orange-yellow with a golden hue, sometimes marked with white or black; the abdomen may be either white or yellowish.

The name of the entire family, of course, had its origin in the prevailing golden color of the species first introduced into Europe. The dorsal fin is long, reaching from the middle of the back nearly to the tail; specimens with a short dorsal fin are deformed. They may sometimes have a divided tail, giving to them, if viewed from behind, the shape of a reversed letter "Y." The size of the goldfish varies according to the locality in which they are raised, and the circumstances surrounding them, it being possible for them to attain a length of eighteen inches and a height of six inches. It is said-that they may live to be a hundred years old, but this may be an exaggeration.

Goldfish enjoy a warm temperature, in bodies of water without a current; just such water as ponds afford suit them best, though they will make themselves at home in streams and multiply exceedingly. They are very hardy, and thrive well under circumstances that would be fatal to many other fish, this quality particularly adapting them for life in the aquarium, specimens having been successfully kept in this way, in good condition and health, from ten to sixteen years. So far as food is concerned, the goldfish will feed upon almost all kinds of vegetable matter, insects, worms, etc., even preying upon small fish, and devouring its own spawn and young. The food is taken in by a sucking motion of the lips; the mouth being toothless, as in all carps, the mastication is accomplished by a few bony tubercles situated in the throat.

In the spring or summer following the one in which they were hatched, goldfish attain their maturity. The size of the fish has nothing whatever to do with the ability to spawn, though a large fish will deposit more eggs than a small one. As an instance, goldfish one inch in length, nine months old, spawned in an aquarium kept in a parlor; all the eggs hatched and the young grew up. In the spring of that year, when the temperature of the water rises above 60° F., they become lively and vivacious, losing all timidity and precaution, so that they easily fall a prey to their many enemies.

Two, three or more male fish follow a female, chasing her to some shallow place where there is an abundance of water-plants.

They lash the water in a lively way, twisting the posterior portion of their bodies energetically and shooting through the water near its surface with short tremulous movements of the fins. At places they gather together in a compact mass, one tumbling over the other. This is the moment when the female drops her eggs, which are immediately impregnated by the males.

This process is repeated throughout the summer, with intervals of rest during the hottest period. The eggs are of the size of a pinhead, and may be either semi-transparent, yellowish, or brilliant yellow in color. Whether this difference in the color of the eggs has anything to do with the coloring the fish will acquire after it is hatched still remains an open question. With the eggs of the trout the case is different, for the culturist can predict the color of the flesh of the fish when grown up; it is possible that this may also apply to the goldfish. The eggs are covered with some adhesive substance, mucus probably, and adhere to anything they happen to touch. The water-plants in the immediate vicinity of the place where the fish have been rolling about will, upon examination, be found covered very profusely with them.

The young are hatched out in from two to six days, the period of incubation being determined by the temperature of the water and the condition of the weather. Direct sunlight has the effect of hastening the process.

During the first few days the young fish are not able to move about much; they hang or lay about among the water-plants, obtaining subsistence from their yolk-bag, but as soon as this is absorbed they swim around on the search of something to eat.

The color of the skin of young goldfish is at first silvery gray, but at an age of six weeks this color begins to change, becoming darker and assuming a cloudy appearance, finally taking on the permanent color. The whole process is sometimes completed in two days, though in some instances it is deferred until the following spring.

The perfection and rapidity of the coloring process depend upon several causes, foremost among which is the proper selection of specimens from which to breed.

This is a very important consideration, in truth, the prime factor upon which hinges the character of the result.

When hatched, the young fish are further subjected to modification by their surroundings, the temperature of the water, its depth, quality, etc., all exercising more or less influence.

The complete result of the breeding, so far as the coloring is concerned, can not be seen until the change has taken place, when any errors that have been made are at once apparent. Those fish that have failed to receive any coloring are then called silver-fish; should they have turned milky white (albinos), they are known as pearl-fish.

The young, when kept in warm ponds—this means that the ponds are so located that the sun can warm them thoroughly—may grow six inches long in four months. As a general thing, however, the length reached in that time is from two and a half to three inches.

In connection with this may be mentioned that all the young of the same spawning do not grow at an equal rate, some few doubling the average size, while others fail to reach it. The latter are then known as dwarf goldfish, and are much in demand for small aquaria.

Deformities occur oftener in this family of fishes than in any other. Specimens minus a dorsal, anal, or even caudal fin, are frequent; rarer are those destitute of scales or minus an eye or with a lob-sided mouth.

From the habit of the goldfish to seek its nourishment on the bottom of the pond in which it lives, and to its careless sluggishness, caused by its voracious appetite, the goldfish, especially in its younger days, falls the victim of innumerable enemies.

In fact, so easy is it to destroy them, that anything else living in its company in the pond and of sufficient strength to master it, may be set down as its enemy.

When arrived at maturity its peculiar mode of reproduction renders it a welcome prey to enemies that do not live in water. It is likewise subjected to diseases of various kinds, and considering all that, it may be said that it is a mere accident when goldfish multiply to any considerable extent without the protection of man, or that the circumstances under which they increased were exceptionally favorable.

# 3 Japanese Goldfishes

*(Carassius japonicus.) Japanische goldfische. Poisson d'Orient.*

In introducing this variety of fish to the attention of the reader, it may be said that the description of the mode of reproduction, habits, etc., regarding the common goldfish, equally apply to the Japanese and Chinese varieties, as they are members of the same family, differing only in shape and color.

The Japanese specify the goldfish by the shape of the body and that of the fins, the coloring in most species being the same. The prevailing colors are vermillion, gold and white, if indeed the last can be called a color.

The Japanese classification being rather complicated, the species' names will be omitted for convenience sake, and the fish described in the manner others have been, though at the same time retaining the foreign nomenclature.

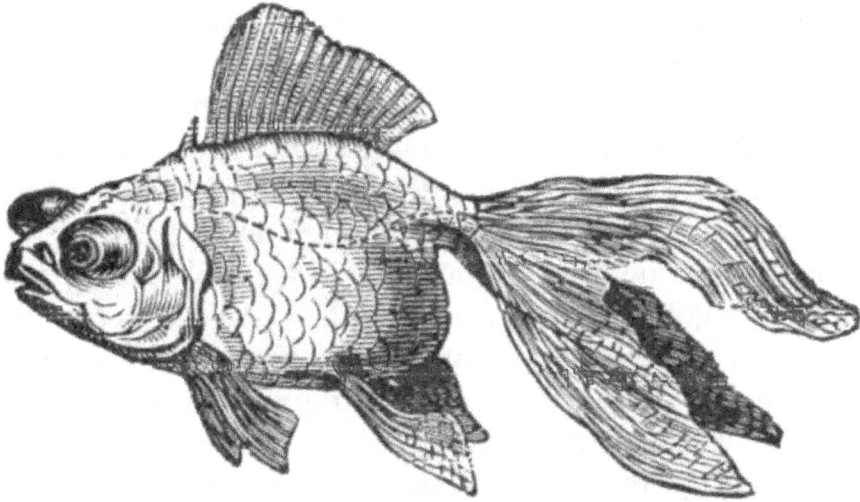

*The telescope*

## The Telescope

   This fish is indeed an odd looking affair. The body is spherical very much like that of a frog-tadpole, and covered, as all gold-fishes are, with medium-sized scales. The fins, are very delicate in structure; the anal fin mostly absent; the caudal fin is double, very large and deeply divided. The eyes are large and project forward, having the appearance of a small telescope adjusted to the eye (see illustration), from this the fish received the name it bears. To make the eyes more prominent the Japanese culturist resorts to an ingenious device. He places the young fish in small dark-glass vessels shaped for the purpose, and which obliges the fish to look constantly in but one direction.

   The color of the fish may be either vermillion, white or part of both, the markings in that case being very beautiful. The body of the fish is all out of proportion to the size of the fins, and in consequence propulsion is a difficult matter.

In spawning, the male rolls the female about among the stones in a most pitiful manner, sometimes for days together; this is an effort of the male to assist in the extrusion of the eggs. When spawned the eggs attach themselves to the stones and other substantial objects rather than to water-plants, probably because it is amongst the stones that the extrusion takes place.

When the young are first hatched, they appear exactly like the common goldfish. They rest upon the water-plants or other supports for a couple of days, at the end of which time the yolk-bag is absorbed; then commences the struggle for existence. The double tail, which is even then large enough to be distinguished, hinders the tiny creatures (but one-quarter of an inch in length) in their movements in the water. Should they find sufficient food to make a bountiful meal, matters become still worse, they lose their balance and can not go at all until digestion is completed, in the meanwhile falling an easy and welcome prey to numerous enemies.

For this reason, which is the main one, and because the fish is very difficult to propagate, it is exceedingly rare, even in Japan. When mature, the telescope fish is about the size of a man's fist.

## The Fringe-Tail

The body of the fringe-tail is short, egg-shaped and slightly compressed; the eyes normal, but very variable in the color of the iris, which is that curtain in the anterior portion of the eye that by its contraction and dilation, regulates the quantity of light that enters the optic. In this fish the color of the iris may be any color, excepting green only, in different individuals.

The fins are large and of very fine structure; the anal fin is double, while the caudal may be either double, treble or quadruple, as the case may be, and is larger than the body, drooping very gracefully. (See illustration.) These special characteristics, among all the fancy varieties, are not fully developed until the second year. In the coloring of the body and fins this fish is not surpassed by any other, making one of the most valuable and desirable objects for the aquarium. There are specimens the back and sides of which are deep vermillion, the abdomen, throat and eyes of rich gold, while all the fins are milk white. Others again will be found presenting a rich vermillion on that part of the body forward of the dorsal fin including the throat, part of the abdomen, and the respective fins, while all back of that is pure white. Some there are with body all white, the fins red, and *vice versa*; others pearly, dotted with irregularly shaped pink spots, the eyes being blue; again, the entire fish will be white, the only touch of color being the deep Ted of the large eyes. In fact, it is impossible within the limits of this sketch to enumerate the endless variety of the markings to be seen, and one can only get an adequate idea by examining a large number.

Strange as it may seem, this beautiful species of fish was nearly last to this country through the caprice of private individuals who happened, or rather were lucky enough, to possess perfect specimens, and were unwilling that others should enjoy the possession of the like. But fortunately, a lady both generous and appreciative, rather than that the stock should die out, loaned to the author for the purpose of reproduction several fine specimens she had recently obtained. The pure stock may now be considered secure for the future. In evidence of the extreme beauty of the species, it may be stated that private parties paid for fine specimens twenty times the weight of the fish in gold.

What has been said about the difficulty of raising telescope fish, may with almost equal propriety be repeated for the fringe-tail. Unlike the former, the latter fish spawns against aquatic plants, the extrusion of the eggs not being so difficult.

*The fringe-tail*

## The Fantail

The body is elongated and compressed on the sides; the head pointed; the fins are short and stout, the anal sometimes double, and occasionally found wanting. The caudal fin is comparatively short, is double, with the upper edges grown together; it is sometimes erect like that of the fantail pigeon (*see frontispiece*), or projects horizontally. The colors are mostly vermillion and white; in some cases the whole body is white, with the exception of the abdomen, this being golden; in others the body is dark red, the belly also golden. The color of the eyes is variable.

It may be mentioned here as an advantage of the double-tailed fish, that they are unable to jump out of the water. This fact is undoubtedly one to be appreciated by those keeping an aquarium.

## The Comet

This is a noble looking fish, and greatly resembles the fantail.

Its body is slender, the fins very large and of fine structure; the caudal fin is single and deeply divided. The coloring is identical with the preceding.

## The Nymph

The body of this species is oblong, and much compressed on the sides; the head is short, and fins normal. The color varies from a light to a deep vermillion, sometimes white; a background of white beautifully dotted with crimson, the throat golden, makes a magnificent combination that is very ornamental indeed.

The sub-varieties of this are the Ramsnose and the Hognose.

## The Ramsnose

That part of the body of this fish from the mouth to the dorsal fin forms a bow, like the forehead of a ram, this feature giving the fish its name. The lower part of the body from the mouth to the tail continues a straight line.

## The Hognose

The peculiarity of this fish is just the reverse, concaved, greatly resembling the head of a fat hog.

All of the goldfish described above do not attain a very great size, seldom attaining a length of more than eight inches.

# 4 Chinese Goldfishes

*(Carassius orientalis.) Chinesische goldfische. Poisson d'Orient.*

The Chinese species of goldfish differ more from each other in color than in the shape of the body and fins, the contrary being the dominant characteristic of the Japanese species, described above.

So startling are these colors that one is almost immediately reminded of the story in the "Arabian Nights' Entertainments" of the enchanted lake, the fish in which were of four colors, yellow, blue, red and white, and were people of four races, who had in some way incurred the displeasure of a genius, who, in revenge, had thus transformed them.

The Chinese reckon seven distinct species, each with its sub-varieties, the most of them very beautiful indeed. Of the species King-Yu, the variety called

## The Mottled Beauty

is very remarkable. Beneath, the fish is simply silvered, but on the back and sides it is mottled with rich hues of blue, yellow, black and rose, the latter color deepening into pure crimson at the gills. Along the ridge of the back, especially near the tail, the black becomes very intense, and has the rich appearance of a piece of Lyons velvet.

Another of the same species is

## The Superb

This magnificent fish, when fully developed, is nearly fifteen inches long. The under part is silvered like the preceding, but the back, which is remarkably broad, is splendidly varied with scarlet and black, the scales being delicately edged with a metallic gold color.

## The Small Blue

This, like the others, is also silvery on the abdomen, but which, however, is delicately flushed with pale rose. The whole of the sides and back are of a rich azure, which shines with a metallic luster.

## The Moor

This is another distinct variety, and is a large fish. The back and sides are very nearly black, which becomes violet underneath. All of the scales are edged with a red-toned bronze.

## The Pie-Bald

*Species Ouen-yu.*
This is a very beautiful variety, the ground color of which is a delicate flesh tint, and upon which there are large patches of rich brown, very much like the markings of a pie-bald horse.

## The Ruby

*Species Nin-eubk-Yu.*
This is a fish of exquisite beauty, being of a delicate semi-opaque, violet crimson, gradually shading off to pale rose underneath. A peculiarly fine variety of this species is

## The Red-Fin

The body of the red-fin is a delicate azure, while the head, tail and fins are an intense scarlet.

## The Tumbler

*Species Kin-teon-Yu*
This is a remarkable fish, and one that seems quite distinct. The head and tail are bent upward, giving to the entire fish the form of a crescent.

When swimming it has the habit of throwing itself over and over in the same manner that tumbler pigeons do when flying. The color of the fish is a magnificent blue flushed with orange making a gorgeous display.

## The Elegant

This is a variety of the species Ouen-Yu, or "lettered kind," so named because the streaky markings are not altogether unlike Chinese characters.

It is of a pure white sparingly dashed with patches of pearly pink on the body, and having some exquisite letter-like markings or tracings about the head and tail. Another variety of the species is rich scarlet, shading off to black on the back, in the midst of which is a large cross of pure white having two transverse bands like the *Cross of Lorraine,* by which name the author thinks proper to designate it.

*Part II*

# THE PONDS AND THEIR CONSTRUCTION

# 5 About Fish-ponds in General

*"Withholding facts is robbery."* —*Orville Dewey.*

A basin of water in the open air, so arranged or constructed that the water may be drawn off at any desired time, is called a pond. A body of water not under control, in the full sense of the word, is usually known as a pool or a lake. The dimension of such body of water does not necessarily regulate the name applied to it, as for instance, a pond may either be but a few feet in length and breadth, or it may cover an estate of hundreds of acres in extent, just as the requirements of the case may be.

In countries far removed from the sea-shore, or any other natural source from which fish may be obtained in large quantities, the necessity arises for supplying the want by some means or other. This can only be done by cultivating them on a large scale, and in quantities sufficient to meet the demands of a steady market. The cultivation of fish for the purpose of food has been carried on for many generations, and so far back as the Middle

Ages considerable attention was given by the monks to pond culture, in order that they might have a reliable source from which to procure fish during lent.

From those days virtually dates the systematic culture of fish in waters that are thoroughly under control.

At that time means of transportation were meager and very slow, so that the cultivation of fish in the interior was a prime necessity, yet notwithstanding the rapid transit on the railroad of the present time, putting inland places in close connection with the ocean and other natural fish reservoirs, pond culture may be and is carried on with both success and profit. Though the railroad has made the saltwater fish a great rival of that Cultivated in fresh water, the latter nevertheless has the advantage of always being marketable and close at hand.

To so successful a degree has pond culture arrived at the present day, that one is enabled to propagate the scaly tribe in quantities without limit.

The fish that was principally, if not to say exclusively, cultivated in ponds in former days, was the famous German carp; the establishments founded three and even four centuries ago being still in successful operation, and may at any time be seen in various parts of Austria and Germany. So thoroughly has the industry been studied that it might almost be dignified with the name of a science. The profits arising from it are enormous, and it may be that interest has had something, if not very much, to do with the pond culture as it is now carried on.

To give the reader an idea of its extent in those countries, it may be said that the carp ponds belonging to the manor of Wittengau number 250, and cover an area of 22,000 acres, the annual yield of fish from which is one-half million of pounds. Similar establishments are found on the manor Konigswartha, in Upper Silesia, with 205 ponds, covering 9,000 acres; the manor Peitz-Cottbus (Brandenburgh) with 72 ponds, covering 5,600 acres.

These and many other large establishments, to say nothing of thousands of ponds scattered all over Central Europe, give ample evidence that the industry is one of magnitude and importance.

In this country the cutting down of timber, the draining of the land, and the establishment of new industries have no doubt greatly increased the value of the soil, but have also influenced the decline in the productiveness of the water.

This is a matter of serious importance, and one which is already claiming the attention of intelligent journalists who see the necessity of fighting the evil ere the remedy is beyond our reach. In fact, the fish industry of the country is one of national interest, and was, not many years ago, the cause of diplomatic difficulty between England and the United States, putting the latter to the expense of millions of money before the matter was satisfactorily adjusted. In view of these facts it will not be long before the necessity of cultivating fish for the market will make itself felt, just as it is in Europe to-day.

The consumption of food-fish, of course, increases with the increase of population, while on the other hand, for reasons given; the supply is rapidly decreasing in quantity and quality. Besides the creeks and rivers are now becoming the sewers into which the washings and dirt of the nation are poured, and the gradual destruction of our fish is consequently taking place right under our eyes.

In the author's opinion, the decrease of our fish is not so much due to the use of small mesh seines, as many pretend, as it is to the increased demand for fish as food. The destruction caused by a few careless fishermen is more than counter-balanced by turtle-hunters and snake-killing boys, the one making it a business, the other engaging in it for sake of the sport, who between them catch and kill a multitude of these voracious enemies of the fish, which, in the water all the time, manage to destroy untold numbers of fish.

The oft-repeated remark, "in former years this river used to be alive with fish," is very well in its way, and is, moreover, true; but it must not be forgotten that in former years there were not so many people requiring them for food.

A bit of forest in which grow a few chestnut-trees was, by the then few village boys said to be "full of chestnuts," but now, when the village has become a large town or city, the few chestnut-trees are not looked upon with the same admiration because, though in equally good condition, they do not furnish chestnuts enough to go around.

A piece of land of given size supplies in plenty, vegetables for one family, but if the family increases in numbers, and additions have to be made to the house at the expense of the piece of land, it can no longer furnish the original quantity of food. New land then is added to the field, or, if that is impossible, that remaining is forced to increased production by the use of manures.

Now then, why not apply to pisciculture the rules guiding agriculture, when viewed in this light? The fish-consuming family has inordinately increased, its dwelling is enlarged at the expense of the fish-producing rivers which are now made to serve other purposes. New additions are consequently necessary and possible by the proper use of water-courses that are now going to waste.

Pond-culture is not solely confined to the production of food-fish, there being many large establishments devoted to the exclusive culture of ornamental fish.

The largest of these are also in Europe, one in Oldenburgh covering twelve acres, yielding annually 300,000 goldfish, and affording employment to many people. A still larger establishment is that in Austria, belonging to Baron de Washington.

There are besides many smaller establishments scattered about all over Europe, and generally near the large cities, just as we find florists in the immediate neighborhood of our own cities.

# 6 The Construction of Ponds

The fish-culturist recognizes three kinds of ponds, which are designated by the source from which they receive their supply.

First, the *creek* and *river ponds;* these are fed from the source that gives them their names, and which may be, in fact, usually is, in the immediate neighborhood. The water may be conveyed to the ponds either by the action of a pump or by means of a drain, the latter undoubtedly being the least expensive, and the most reliable, besides requiring no further attention than to turn on or off when so desired.

Second, *spring-ponds,* those fed by a spring, and it may be said in favor of such ponds that the supply of water is steady and of uniform quality and temperature, besides being free from mud.

Third *sky-ponds,* those that receive their supply from the rain that falls, and which is drained off or runs from the surface of adjoining fields.

The value of a pond is based upon the reliability of its water supply, the quality of the same, as also that of the soil that forms the floor of the pond. Last but not least, is the location of the pond. The locality ought to be such that without any special expenditure of time, the pond can always be under supervision, thus guarding against pilfering upon the part of mischievous boys, the depredations of muskrats, herons, cranes, etc. The water that supplies the ponds would better come from a spring or be gathered in the manner described for the sky-pond, as by this way one will avoid the annoyance and destruction caused by snakes, muskrats, frogs, etc., that are always found in small rivers and creeks, and which are sure to find their way to the pond, causing much trouble, to say nothing of the money value of that which they destroy. Besides this, the ponds should have some natural protection against the high winds of spring time. This is very well afforded by a clump of trees close by, but much better if the pond is situated in a hollow between two or three small hills. These not only shelter the ponds, but tend to keep them warm by retaining the warmth of the sun in the spring, just at a time when it is most needed. In a warm and protected location of this kind, the fish will spawn several weeks before the ordinary beginning of the season, the young, of course, being ready for sale that much sooner, and the ponds are vacated for a second crop, thus bringing a quicker and more liberal return on the investment.

Lastly, it is of but little use to locate ponds in a neighborhood that does not afford a market for the product, unless, indeed, the amateur should be in correspondence with such parties as would be likely to find it to their interest to handle his stock.

The foregoing remarks on ponds and those which follow are made with an especial view to the culture of goldfish, so that the novice as well as the amateur may get a distinct idea of the proper method of locating and constructing them, for upon this depends the success of the enterprise, and more so when it is engaged in for the purpose of yielding an income.

We come now to a consideration of the immediate and practical points concerning the subject upon which we are engaged
and which we wish to see in running order. After having determined to devote his spare time to the cultivation of goldfish, the
amateur will hardly know where to commence or what to do
first, neither will he know how to do it. That want the author
thoroughly appreciates, and it is his aim to supply it in these
pages, at the same time bearing in mind that the great majority
will probably have but a small capital with which to commence
business, and must necessarily confine themselves to such things
as are most needed when making a start. After a beginning has
been successfully accomplished, other and useful accessories
may be added from time to time that will undoubtedly greatly
reduce labor and enhance the enjoyment of the pursuit, for if
one does not take any interest in what he is engaged he can not
hope to derive from it either pleasure or profit.

The first thing then is to determine on a favorable site for the
location of the ponds. This, as previously stated, should be one
protected from cold winds and fully exposed to the sun, not
forgetting that the soil to form the floor of the ponds should
be as rich as possible. It must also be borne in mind for it is
a very essential requisite, that every pond must be drawn off
at times, and independently of all others. When this can not
be accomplished by turning to account the natural declivity or
resources of the site, the series must be so arranged that this
may be accomplished in sections, or, in other words, the ponds
so situated that two or three or more, as the case may be, can
be emptied at will.

The manner in which ponds are to be constructed in any
given case is, as a matter of necessity, to be governed by the
circumstances as they exist, anything favorable for them to be
taken advantage of and turned to good account. A natural valley

or hollow may be dammed up at the lower end, thus enabling the culturist to place his ponds on top of the ground, so to speak, or they may be made by digging out the soil to the required size and depth to receive the frames hereafter to be described.

As before stated, there are many enemies which prey upon the goldfish, and all of which are to be guarded against in the construction of the ponds in which they are expected to live and increase. If left alone and without the protection of man, the circumstances must be exceptionally favorable under which they can thrive and multiply. For the systematic raising of goldfish, therefore, a series of ponds is indispensable, each of which is constructed and arranged to meet the requirements of the specific purpose for which it is intended. All of them, however, are based upon the same fundamental principle and must have adjustable inlets and outlets, otherwise they can not be under complete control, the reason for which a little experience will soon explain. The number of ponds needed for the culture of goldfish does not exceed four, and in their order are the spawning-pond, rearing-pond, storage-pond, and winter-pond. The *spawning-ponds* or "beds" are those ponds in which the parent fish are placed when the breeding time arrives, and are to be kept there throughout the duration of that season, and solely for the purpose of depositing the eggs. The *rearing-ponds* are those in which the young fry are placed or "planted" and to keep them until they are of a marketable size. The *storage-ponds* are for the purpose of keeping saleable stock until wanted, and in a situation readily accessible at a moment's notice. The last in order is the *winter-pond, in* which the parent fish, and those of the younger that are intended for the same purpose, are carried through the inclemencies of the winter.

The greatest depth of all the ponds, the winter-pond excepted, should not be made to exceed twenty-four (24) inches, and this

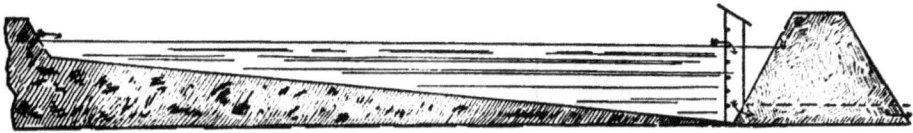

*Longitudinal section of goldfish pond*

to be at that end where the outlet is placed. From twenty-four inches at one end the depth should gradually decrease until the water depth at the head is not greater than six (6) inches. (See illustration.)

The upper edges of the dams of the ponds should not be less than six inches above the highest water capacity, thus avoiding overflow during heavy rain-storms; neither should they be less than eighteen (a) in width. It is also necessary that the dams be made very carefully, so that when soaked with water they will not sink or cave in with their own weight. A most excellent material of which to build them consists simply of sod cut into suitable pieces and laid one upon another, just as a stone-mason lays one stone upon the other, the whole when thus laid becomes very solid and compact, and capable of resisting considerable pressure, and will withstand the wear and tear of the weather and the weight of the body when walking over them. Where there is danger prevailing from a sudden freshet in an adjoining creek, it is a wise precaution to build, in addition, a strong dam on the outside and in the direction from which the danger is expected. For greater safety this dam may be in its turn protected with boards, so as to prevent gradual washing away. Each inlet, *e.,* the point at which the water supply makes its way into the pond, is to be guarded with galvanized iron netting of a tolerably coarse mesh, about eight to the inch, as a small mesh clogs very readily, and only adds the additional care of keeping it clean. The outlet is formed of two gutters of which one fits in an upright

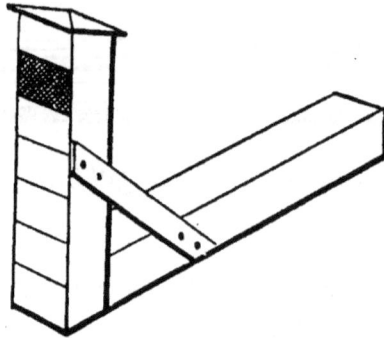

*Outlet for pond*

position tightly on the end of the other, which leads through the bottom of the dam, forming a right angle. (See illustration.)

The one leading through the dam is closed tightly on all four sides, but on the upright one the side facing the pond is closed with adjustable sections cut from flooring boards. By means of these movable parts the level of the water in the pond can be regulated as occasion may require. The overflow, that is, where the water is running out, is guarded by a wire screen of the same sized mesh as before mentioned, and adjustable in the same manner as the other boards protecting the opening. When it is desirable to drain the pond, one section or board is removed at a time, always taking care to place the wire screen upon the top by letting it slide down into the grooves. (See illustration.)

A most important point, and one to be insisted upon in the construction of ponds, is to so arrange the water supply *that it will at all times be under complete control.* Each pond is to receive its supply independently of all the rest, and the water must also, when need be, be cut off from the pond without interfering with the supply of others. The supply channel must likewise have an independent "run," so that the water it contains when not wanted for use in the ponds, may find its way out of the establishment. This point is one of great importance, especially during heavy rain-storms, as the large increase in the volume of water would cause the ponds to overflow and probably do great damage.

# 7 The Author's Establishment

The accompanying diagram will serve as a reliable guide for the construction and arrangement of a complete establishment for the raising of goldfish.

It will be observed that the ponds are arranged in a semi-circle, not because the "horseshoe" is all the rage, but simply for the season that the little valley in which they are located is encircled by small but steep hills running in that direction. As previously stated, the disposition of the ponds must of necessity be in accordance with the topography or "lay of the land" in the locality determined upon, and the amateur will have to follow accordingly, though there are changes and modifications that he can make, and in many instances with great advantage.

## DESCRIPTION OF PLAN

A.—Spawning or propagating beds.
B.—Rearing ponds.

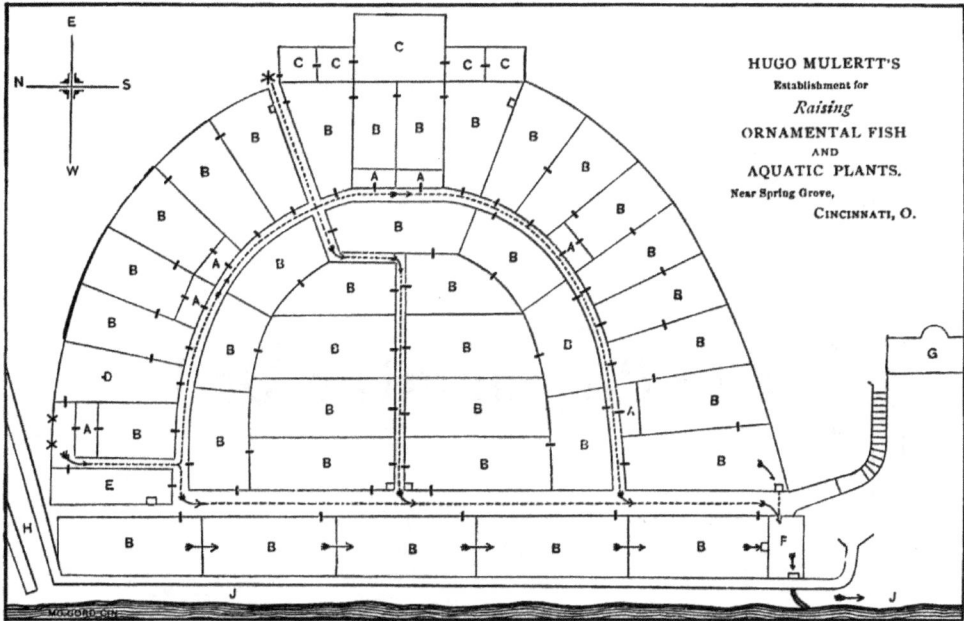

*Establishment for raising ornamental fish and aquatic plants*

C.—Storage ponds and nursery.

D.—Winter pond.

E.—Isolated pond for special purposes.

F.—Catch-basin for stray fish.

G.—Lodge, where the eggs are hatched.

H.—Wind-brake and guard against freshets.

J.—Creek, passing the establishment.

Star.—Location of springs.

Arrow.—Direction of water flow.

Dotted lines.—Supply drain.

Bracket.—Outlets.

# 8  The Propagating Bed

Having now had a general view of the ponds, the reader's attention is directed to a detailed study of them, each one separately, the method of their construction, and the reason why they differ from one another; for, being made—each with a specific purpose to serve, they of course must be built with that object in view.

The spawning-pond, or "propagating bed," is situated at that end of the rearing-pond where the water makes its entrance. It is a division consisting of a board frame eighteen (18) inches in depth and four (4) feet wide by eight (8) feet long, fashioned very much like the frame of a gardener's hot-bed. This frame, when ready, is to be sunk about two inches into the soil composing the bottom of the pond, and very accurately at that; for if this precaution is not taken, the fish are very apt to escape from the pond through any aperture that may be left. The upper edge of the frame should be not less than six inches above the usual level of the water, thus preventing the danger of overflow. In the side facing the supply drain, about two inches above the water level, and in the other forming the division of the rearing-pond, about one inch *below* this level, small openings for the entrance and exit of the water are cut and carefully covered with galvanized iron-wire gauze, of about eight meshes to the inch.

This box, the bottom of which is formed by the floor of the pond, is to be covered with a well-fitting frame, mounted upon hinges and also covered with galvanized iron-wire netting of one mesh to the inch. This wire screen serves to prevent the approach of numerous enemies that would otherwise greatly interfere with the spawners, if not destroy them altogether. By following the above given directions for the construction of the frame the greatest economy possible can be had, for the lumber in the board as obtained from, the dealers, measures twelve (1 2) and sixteen (16) feet in length, so that the measurements, eight by four feet,. can always be had without unnecessary waste of lumber. Besides this, the wire netting is obtained in any length, and also four (4) feet in width, so that here too is a saving of material, and consequently of expense.

Inside of this frame, or box, as it may be called when placed into position, another frame upon which wire netting is stretched, is to be sunk. The size of the mesh is to be such as is best adapted to the size of the fish that are to be placed in the pond. It is very easy to see that this false bottom subserves a very useful purpose, and often saves time that at the moment is otherwise valuable. When, for one reason or another, it is desirable to remove the fish, the only requirement is simply to raise the frame to get it into shallow water, when the whole contents of the pond can be examined at ease.

A spawning-pond, such as the one described, can be, and with propriety too, constructed separately and wholly isolated. But if it is made a division of the rearing-pond, a good advantage is secured, namely, the stream of water that passes through it will carry into the rearing-pond such of the young fish as were hatched from undiscovered eggs. They will thus escape being devoured by the parent, and besides will get into the place intended for them, and where the chances for their growing up are vastly in their favor.

## The Nursery

This is merely an apartment temporarily established in the storage-pond, which serves the purpose of protecting the young during their earliest infancy.

## The Rearing Pond

This pond differs from the preceding in several respects. To begin with, it is five times as large, the size eight (8) feet by twenty (20) being in many ways the most convenient, as experience has abundantly proven.

In constructing this, as well as the others, it is not absolutely necessary that the outlet should be directly opposite the inlet, though such an arrangement insures the most complete changing of the contents (the water) by the current passing from one end to the other. The location of the outlet and inlet must be determined by the judgment of the builder when seeking to get the best results he can, from the circumstances and surroundings of the locality in which he places his ponds. It is necessary, however, that the inlet should be one or two inches above the water level, so that the fish will be prevented from getting into the supply drain, should there happen to be a defect in the guard. In constructing the pond, the remarks applying to ponds in general are to be respected. The dams must reach six inches above the water level, and the floor be as uniform as possible, with a regular incline of from six (6) inches depth at one end to twenty-four (24) inches at the other. The deepest part to be at the outlet, insuring a perfect and even drawing off, whenever that may be needed.

One or two plants, such as Nymphaea only, should be planted in the soil of the bottom, as their floating leaves afford shelter to the fish in very hot weather. Besides it offers greater convenience for the fish to reach deep water upon the approach of danger, as in such cases fish always dart to the bottom, and being there, they get out of the way much sooner. If placed near the outlet, it is also easier to gather the flowers when the plants are in blossom; this is probably a small point, but one that will be appreciated upon trial. Should the plants not be available, one or two pieces of board left floating on the water will answer the purpose. If it is impossible to isolate the rearing-ponds from each other, the consequences are not at all serious, but things should be so arranged that the very large fish are not mixed up with the small ones, as the latter would then be deprived of their share of the natural food the pond itself supplies.

## The Storage Pond

This pond is for the sole purpose of keeping the saleable fish where they can easily be secured when desired, assorted according to variety, color and size, and kept in good condition till wanted.

It is to be so located that it will not interfere with the draining off of the other ponds.

Its construction does not differ in anywise from that of the rearing-pond, as indeed it can be used for rearing, if not wanted for nursery purposes (of which we will treat further on), until wanted for its destined purpose.

It is best, however, to divide it with boards into different parts of equal size, so as to secure a general fit of wire covers.

## The Winter Pond

This is best located close by a spring, so that the constant flowing in of a stream of water of even temperature will prevent the surface of the pond from, freezing over entirely, in which case the fish would die. The depth of the pond ought not to be less than three feet, and its sides closely lined with boards or masonry, thus affording protection from enemies and preventing caving in. The flow of water is to enter at one end, traverse the length of the pond, and leave at the opposite extremity. The dimensions of the winter quarters need not be very large, as during the winter season, fish in a natural state, pass that time in a semi-torpid condition, eat no food, and are not revived until they get the warmth of the sun in early spring.

A compartment of four (4) feet by eight (8), and of the above depth, will carry about fifty fish of eight to ten inches in length safely through the winter.

Where it can be done, the fish may be wintered over in a green house cistern, or in aquaria where they may be under close observation, and at the same time be of some ornamental value. If in the open air, the winter-pond is, of course, to be covered with wire netting to keep out the various birds and animals that would prey upon them; it is also well to protect the greater part of the top with boards, keeping out as much snow as possible.

*Part III*

# The Propagation and Care of the Goldfish

# 9 The Anatomy of the Goldfish

*"Nature my school, the water my field."*

Before we enter into the details of the propagation of the fish, it is necessary to learn something about its anatomy.

A study of the accompanying cut (which is merely diagramatic) will greatly assist the reader in getting a clear understanding of the internal arrangement of the several parts with which it is most necessary to be familiar.

We will consider first the *scales,* as they are almost the first thing to strike the eye. These scales are so arranged upon the surface of the body that they overlap each other just in the manner that a carpenter lays shingles on a roof, being disposed in such a way that the friction incident upon, the movement of the fish in the water is reduced to the smallest proportion. As the fish grows older and larger the scales increase in size. Near the middle of the body and running along each side of the fish, there is a line or row of scales that possess peculiarities distinguishing them from other scales. These scales are pierced with a tubular aperture, and the tubes of which they are the exit are quite distinct, and form the so-called "lateral line." Through these tubes a slimy substance or mucous is exuded, which covers the entire body, seemingly for the purpose of making the fish waterproof and of further reducing the friction in the water. These tubes always point from the matrix, or root of the scale, towards the tail of the fish. It is scales of this description that naturalists refer to when seeking to learn the species to which the fish belongs, because the peculiarities of their structure differ in them. (See illustration.)

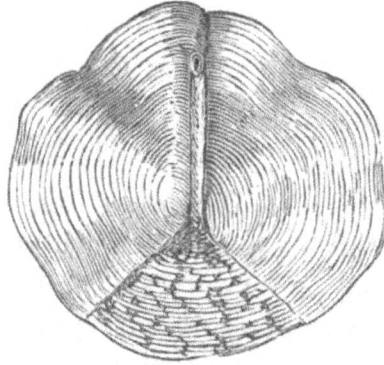

*Scale from lateral line (enlarged)*

The next thing most noticeable are the *fins*, these being named according to their location upon the body of the fish and subserve various purposes. The *pectoral* fins (fig. a.) are those situated in the place corresponding with the fore-legs of animals, the *ventral* fins (fig. b.) being placed where the hind-legs of animals are found. The *dorsal* fin (fig. c.) is that one found upon the back of the fish. That fin situated behind the anus receives its name from that part, and is known as the *anal* fin (fig d.), while the tail of a fish is properly called the *caudal* fin, (fig. e.)

The water in which the fish lives is very nearly as heavy as the fish itself, the latter then requiring comparatively little strength to move about. The motion necessitating the greatest expenditure of power is that of propulsion forward, and is accomplished by the action of the caudal fin. The pectoral fins are used to change the water in the neighborhood of the gills, thus serving as adjunct respiratory organs.

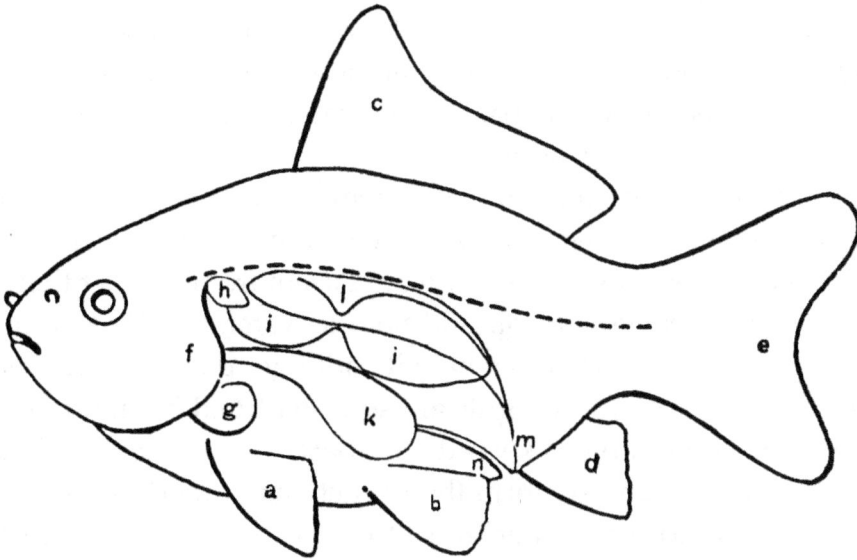

*Anatomical sketch of goldfish*

The ventral fins are mainly useful as a brake when the fish wishes to come to a sudden stop when in motion, or for a backward movement; the dorsal and anal fins serve the purpose of balancing the body.

The *gills*, which are organs of respiration, are situated on each side of the head, protected by "opercles" or gill-covers. (fig. f.) The gills themselves consist of bony arches which are covered with a tissue containing a large number of blood-vessels. As a rule there are four of these arches on each side of the head. The life-giving principle that supports animal life is a gas called oxygen, and as this exists in the water, the function of the gills is to extract it. When the water passes through them, as it does when taken in by the mouth of the fish, and is pressed through the gills, it comes in contact with the blood-vessels, the oxygen being absorbed into the blood for the nourishment of that fluid and the body generally.

It will thus be seen that it is not the water that the fish breathes, as many suppose, but the air contained in it, as can be proved by placing fish in water from which the air has been taken, by prolonged boiling or otherwise.

Fish that are kept in a vessel will come up to the surface where the air can be mixed with the water when that in the water of the vessel has become exhausted. It would seem from the fact that as the fish breathes only the air and not the water, that it could just as well live in the open air, where it could get plenty. Such is not the case, however, for the gills are so constituted by nature that they need something to keep them apart, so that their surfaces may be exposed and perform their functions properly, otherwise they would close together, the blood-vessels would cease to absorb the oxygen, resulting, of course, in the death of the fish. It is necessary, then, that a stream of water should constantly flow through them, as it does, the absorption of oxygen then going on as nature intended.

The *heart* of the fish lies just behind the head and between the gills (fig. g.). It is a muscular organ consisting of three parts, an auricle, a ventricle, and an arterial bulb. The venous or stale blood is pumped into the gills by the heart, where it receives a fresh supply of oxygen. From the gills it is sent to an arterial trunk, lying along the under side of the vertebral column, (fig. h.) from which it is distributed all over the body of the fish.

As fish have no lungs, they can not possess a *voice*.

The goldfish is supplied with a divided *air-bladder* (fig, i. i.) which can be filled or emptied at will. This bladder is a sac formed of a tough membrane, and is situated between the spinal column and the *stomach*. (fig. k.)

It appears that the air-bladder is either for the purpose of increasing the weight of the fish when empty, and decreasing it when full, thus exercising a modifying influence upon the weight of the fish when compared with that of the water.

The *eyes* of the goldfish are well developed, but so far as *hearing* is concerned, opinions differ, and the question is still disputed. It may be stated that if fish hear at all, it is with great difficulty.

It is the custom in some places where fish are kept to call them to their feeding place by the tolling of a bell, and they come, but it *is* a question whether they come because they hear the *sound*, or that they see the *motion* of the person ringing the bell, or that of the bell itself; this, then, can not be cited in proof of the theory that they hear.

Music or the report of firearms does not affect them at all, but the flash from the discharged gun will scare them.

Do fish sleep? Yes.

In the act of sleeping they do not close the eyelids, for the very good reason that they have none; neither do they select the night for the purpose. Goldfish have been seen asleep in the broad sunlight of the forenoon, and the same varieties have also been found sleeping at midnight. Their time for resting then, does not occur at stated periods, but whenever the desire comes upon them. The fish may very easily be observed in slumber, remaining perfectly still, the only motion being that of the breathing apparatus and the pectoral fins, the action being very slow but regular. All the other fins are at rest, and the pupils of the eyes appear to be drawn back. The other senses *taste* and *smell* are very well developed.

We come now to the consideration of the process of *reproduction*. The organs for this purpose are distributed between two individuals—the male and the female. The *female organs* consist of two sacs located immediately below the spinal column on each side of the air bladder, uniting towards the posterior end in a single oviduct which discharges outside, behind the anus (fig l.). These contain the eggs by thousands and which increase in size when the spawning season arrives, greatly distending the ovarian sacs. The ripe egg when it separates from the ovary, passes through a tube (*the oviduct*) (fig. m,) the opening of which, as before stated, is outside, immediately behind the anus. (fig. n.)

In the *male* fish the spermatic organs which are located the same as the sacs in the female, secrete a thick white fluid which contains innumerable small organic bodies, which when discharged and in a fresh condition, move about, enter the egg, impregnate it and start the development of the embryo. These bodies which are called spermatazoa, consist of an anterior thicker part, the so-called head, and the more attenuated part or tail. In the water these little organisms can live but one or two minutes, but when taken from the fish and placed in a bottle kept at a proper temperature, they may be preserved alive for six days. This discovery was made by a Russian fish culturist in 1856 and is very important to the artificial propagation of fish as it enables the crossing of different species.

The *eggs* when first spent in the water have the shape of a slightly pressed in rubber ball, and as soon as they come in contact with liquid, they expand and suck it in through a microscopically small hole. (See illustration.) *The spermatic germs of the male being present in this liquid are thus introduced and fertilize the egg.*

The *fecundation* of the egg consists in the entry of the spermatic corpuscles and the subsequent production of a subdivision of

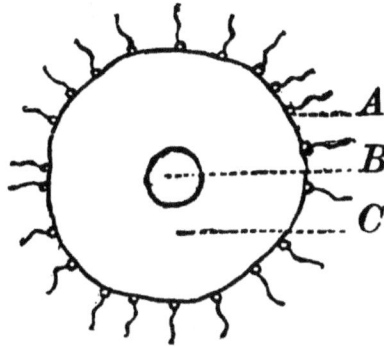

*Fertilization of fish egg. A: spermatic corpuscle. B: germinative disk. C: nutritive yolk*

the germinative din, which phenomenon is called the process of "segmentation "or "furrowing." This is followed by a series of successive changes, of which the final result is the *embryo*, which, subsisting or being nourished by the yolk, gradually develops into the *perfect fish*.

The young fish when first hatched is supplied with a sac called the *yolk-bag*, from which it derives its nourishment during the early period of its independent existence. When this has been exhausted it is then ready to seek other food and this it finds in various microscopic organisms that exist in profuse abundance in the water. As the fish grows larger and gains strength, other and coarser food is sought and devoured.

# 10 Propagation

Among fish culturists it is the universal custom to divide fishes into two classes according to the season in which they spawn; we thus have summer spawning and winter spawning fish. They are further more distinguished as they differ in the mode of depositing the eggs, as some are laid on or in receptacles usually called "nests," especially prepared by the fish for that purpose, and others again are dropped loosely into the water without any precautions having been taken for their protection. In the latter case they again differ in being either "adhesive "or "non-adhesive," in the one instance adhering to anything they may happen to touch and remaining until hatched, in the other sinking to the bottom or floating about at random on the surface.

The goldfish belongs to the summer spawning class, builds no nest and its eggs are adhesive in the full sense of the word. In those fish depositing non-adhesive eggs, the extrusion of Them by hand, and their artificial impregnation is profitable, but with the carp-like fishes, a higher percentage of young is obtained when the spawning is allowed to proceed in its natural manner. Besides, their eggs not being mature all at the same time, would make the operation of extrusion an oft-to-be repeated affair, and which, furthermore, would greatly endanger the life of the fish and require much time and attention.

Goldfish raising is therefore confined to:

1. Guiding Nature with regard to the "survival of the fittest," and with it of course the pure strain.

2. Assisting Nature by securing suitable spawning resorts.

3. Regulating the spawning season, and

4. Protecting their spawn and young.

## Selection of Breeders

As soon as the spring weather sets in with sufficient sunshine to affect the temperature of the water, the fish in their winter quarters rise near the surface and become lively again. This is the time to get the spawning beds ready for action. Whatever month or date that may be, is determined by the respective locality of the ponds, viz., their situation in a northern or southern climate, and there, whether they are exposed or protected. The beds are then filled with water to the proper level, and all details concerning the pond are attended to, so that it will be in complete running order. When all is ready, the fish from which it is intended to breed are selected, and right here reside the fundamental conditions upon which depend the production of a good and saleable crop of fish. The fish used to breed from should be healthy in every respect, of good shape and color, and of gentle, fully domesticated habits. All these qualities will be inherited by their young. In regard to the color of the fish, it is of great importance to know at what age this was acquired, as such fish that colored at an age of six to eight weeks transmit the same tendency to their young on an average of 98%. In contrast with this, those fish whose coloring was delayed until the second year, when bred, produce but 5% of young that will color in the first year, while the remainder do not assume their red, yellow, or white coloring until the second year, a great many never changing, always remaining "silver "fish.

The coloring attained by the fish generally remains so, though there are instances in which the red markings may become milky white, and what was previously white changes to red, or black spots may appear, or if present, be lost. This may occur either wholly or in part. A satisfactory reason for this phenomenon can not be given.

The selection of faultless beauties for breeding purposes, however, is not absolutely necessary, such specimens are best kept for exposition purposes. Any fish whose fins may have become injured in any way, by accident or otherwise, but have grown again in some crooked or objectionable shape, are nevertheless perfectly fit for the spawning bed, if they are otherwise in perfect condition, though for, ornamental purpose they would be of little account.

It may be remarked in this connection, and with propriety, that certain peculiarities in the shape of foreign fish, *e.,* those newly introduced, are in the course of time, lost, when imported into the United States, the change of climate, locality, food, etc., producing gradual changes in their typical forms, assuming or acquiring, so to speak, an American type. In view of this, the culturist should never neglect to infuse new blood into his stock whenever a favorable opportunity offers, for by so doing he can keep it up as near to the original standard as it is possible to do under the change of circumstances.

At the breeding season the sexual differences are plainly re-
vealed to the practiced eye of the patient observer—not before.
Upon close inspection the bony plates that cover the gills, the
gill-covers, or *opereles,* will be found covered with small white
prominences, usually denominated tubercles. *Those fish bearing*
*this distinguishing mark are male fish.* These tubercles appear on
the fish when it is in condition for reproduction, and disappear
when that function ceases to be in an active state. This period
may be of longer or shorter duration in different individuals,
and it will also be seen that the number of the tubercles will vary
in the different specimens upon which they are observed. (See
illustration.) The usual method of distinguishing the male from
the female by noting the presence of a short dorsal fin is not
reliable; in fact, it is not only misleading, but false, as such short,
or more properly speaking, "deformed" dorsal fin, is found just
as frequently upon female fishes. The tubercles are sharp, very
similar to the projections upon a rasp, and seem to have for their
function the assisting of the female to pass its eggs through the
canal. This theory is is apparently substantiated by the fact that
the male uses them in such a manner by pressing against the
belly of the female that one is irresistibly led to the conclusion
that they can exist for no other purpose.

The females to be selected must show an expanded belly,
which evidences the maturity of the eggs, as it has been stated
before, that when arrived at that period, they increase in size,
and it is by this appearance only that the culturist can decide
that the time for the female to spawn has arrived.

It is worse than useless to place in the spawning bed fish that
are not in perfect condition to perform the functions, as some
goldfish are sterile, and will only disturb the arrangements that
have been so carefully made for that purpose. Whether these
particular fish will remain sterile for a season or for as long

*Head of male goldfish*

as they live can not be decided. As a rule, it is best to select three females and four Males when of good size, or six females and four males, when the former are small yearlings, for each spawning bed. It is also important to match the sizes of the fish; should this not be possible, in, case the females are larger than the males, two or three males may be required to mate with the females. The best age for spawners is that between two and four years.

The selection having been made, all those fish not wanted for present use are returned to the winter pond.

# 11 Preparing the Spawning Bed

As the goldfish deposits its spawn upon plants that live in the water, it is necessary that the natural condition of things be closely imitated in preparing the bed in which the fish are expected to spawn. These plants can be obtained from any neighboring creek or marsh (that with a gravel bottom preferred). Those marsh plants possessing fibrous roots are either pulled or dug up with their roots entire; they are then washed thoroughly to cleanse them of the adhering mud, and closely examined to see that there are no eggs of other fish or insect larvae upon them. Having secured enough of these for the present, say half a dozen good clumps for each bed, they are then placed loosely in the water of the bed, along that side where the sun shines upon them in the morning. On these roots the females will deposit their eggs. Later in the season, when aquatic plants have commenced to grow, these roots can be removed and replaced with such aquatics as the Horn or Waxworth (*Ceratophyllum demersum*), and the Canal Pest (*Anacharis canadensis*). These plants are especially good to catch the eggs when dropped by the fish.

The water supply is now shut off from the bed, only an occasional supply being let in to preserve the proper height at which the water should remain.

If the goldfish are kept in places where it is impossible to get them out, and their spawn is wanted, large bunches of the fibrous roots can be tied to a string fastened to the shore, and allowed to float about upon the water. When containing spawn, they can be brought out merely by pulling the string, when the eggs can be removed and taken care of by the one in charge.

When the fish are wintered over in sheltered localities, as a greenhouse for instance, the spawning season will commence a great deal in advance of that outside. In such a location, spawn may be expected from the middle of February, while that event seldom takes place in the open air earlier than the latter part of April or the beginning of May. When the spawning has begun, it is continued until the setting in of frost, with now and then an interval of a week or two.

In northern climes, where the season is short, it may be lengthened by placing over the spawning bed and nursery frames covered with glass, the principle being the same as that of the gardener's hot-bed. If this is done, it must not be forgotten that in fair weather an abundance of air should be admitted, and also that the bed is to be protected from great and sudden changes of temperature, as either would be fatal; in fact, the breeder is to exercise his intelligence in the matter, for he certainly must know that as he is keeping up a kind of artificial climate, he must not forget to preserve it as evenly as possible, otherwise one cold night would (if the frame had not been replaced after airing the bed) destroy many days of patient toil.

If, in between spawning periods, the fish should rest longer than is desired by the breeder, the males are to be changed from one bed into the other, and the water in the latter well aerated by letting a stream flow through it for a whole day, when the spawning will soon be in full process again.

## 12 Care of the Eggs

"In the morning sow thy seed," says Solomon:—This advice of. the Sage of the East has many applications to the affairs of every day life, and amongst other things it may well apply to the subject under consideration, not in its literal sense, but in the spirit of the admonition. As the early morning is the best time to sow seed, so is the morning of the season the best time to spawn the fish, as they are then in their best condition for that process, and besides the young will have a much better opportunity to grow without molestation, their most inveterate enemies not appearing until after the season is further advanced. Thus it is that a greater percentage survives, which, becoming marketable before the main crop arrives, bring higher prices and give the culturist encouragement by a quick return upon his labor and capital.

Again our quotation hits the mark, for the earlier the spawn is taken from the bed the greater will be the reduction of losses, as by leaving it exposed, the spawners themselves devour it.

The early morning is the favorite time for the goldfish to spawn, though it is sometimes kept up until noon. The fish are seen chasing each other and rolling over the material thrown into the water for that purpose. Upon inspection of the loosely floating clumps of roots, we discover that they have adhering to them a great many small round watery-white, creamy or yellow colored balls about the size of a pinhead; these are the eggs of the goldfish. The bunches of roots are then carefully removed from the water, and the individual rootlets bearing the eggs are either cut off with a knife or pair of scissors, or they may be detached with the thumb-nail. Great care must be taken not to disturb the eggs or injure them in any way while detaching the rootlets. They are then placed in a one-gallon candy jar, filled with clear water of the same temperature as that of the bed. Such a jar is best not over-crowded, about one hundred eggs being as many as that capacity can safely and conveniently carry, the object being to give the young fish, when hatched out, plenty of room, both to move about and obtain sufficient fresh water for respiration. When the jars have received their quota of eggs, they are taken into the house or put into some other convenient place selected for the purpose, but in such a situation that they will constantly be under close supervision.

It is best to set them near a window, within the reach of the morning sun, there to remain untouched until the eggs are hatched.

The time required for the hatching varies from two (2) to six (6) days, it taking place most rapidly in warm weather. The temperature of the water, most advantageous for the hatching is between 60° F., and 80° F., more or less is dangerous.

This method of caring for the eggs secures to them a more effective guard against enemies, as well as muddy water, heavy rains, and hail storms, all of which would militate against them if hatched in the open air and in the ponds. The candy-jar system furthermore recommends itself for the hatching of the eggs in this, that the whole process is under complete control and offers every facility for close inspection at all times. The jars can be obtained anywhere, are cheap, and are very convenient to handle, in short, just the thing for the purpose. For convenience of study, the marking of dates, names of varieties, etc., together with any notes that it is desired to make, a piece of paper can be pasted upon the outside of the jar, it will always be there, and the record kept upon it can always be seen at a glance.

# 13 First Care of the Young

About the second or third day after the young have left the eggs, they become strong enough to swim freely about in the water, and can then be transferred to the nursery, temporarily established in some part of the storage pond, which is generally not in use at this season. Here they are carefully guarded until they become about half an inch in length. For this purpose the storage pond is filled with water to a height of six inches at the deepest part; all living creatures, especially insects and their larvae, are to be removed with the aid of a fine dip-net. No plants should be placed in the nursery, as it is important that it always be under the control of the eye. Each morning the young are carefully inspected, and any enemies that may happen to have made their appearance must instantly be removed and killed. During the remainder of the time, a cover is kept over the frame in order to prevent the dragon-flies from depositing their eggs in the water, as these, when hatched, are very destructive to the young fish.

As the different lots of eggs are hatched in the jars, and the young become strong enough to swim about, they are placed in the nursery until a division of four by four (4 by 4) feet contains from four hundred (400) to five hundred (500) fish. The next division is then prepared and stocked in the manner described, and so on until all the spawn has been so treated.

At the expiration of about a week from the time the first young were placed in the nursery, the strongest are taken out in the following manner: A candy jar is filled half-full of water from the nursery, and set right in the center of the bed and resting upon the bottom of it. The largest' fish are now slowly and gently caught with a small dip-net, one at a time, and immediately in the same careful manner placed in the jar, counting them as they are transferred. It is not advisable to place more than fifty of these little fish in the jar while removing them, neither should they remain in it longer than is necessary to transfer them into the rearing ponds, the whole time consumed in the operation should not exceed a few minutes.

The morning is the proper time of day to perform this operation, as the temperature of the water in the various ponds is at that time most uniform. If the manipulation is done at any other period of the day, it becomes necessary to gradually equalize the temperature, as a difference of a few degrees only would prove fatal to the tender young if suddenly removed from one water to another.

# 14 Setting Out the Young Fish

The day before the fish are transplanted from the nursery into the rearing pond, the water is let into the latter, but not sooner than this. The reason for this is, that the water, if allowed to stand longer, would produce an over abundance of food, and with it, of course, a host of dangerous insects. If the young fish were put in the pond, while in this condition, the insects would at once give chase and devour them.

The little food these tiny fish require, during their first days of existence, is found in sufficient quantities in the new water. By the time their enemies make their appearance, especially those of the insect kind, the fish have grown too strong for them, and are very well able to look out for themselves.

In transferring the young from the jar, to the rearing pond, it is much better to sink the jar and allow the fish to make their escape at will. This precaution enables them to gradually become accustomed to the change, thus avoiding a shock by the sudden emptying of the jar.

In this manner the young are "planted "in the rearing ponds, at the rate of 250 to each pond, of 8x20 feet in size. Here they remain until they have acquired their coloring, and have grown large enough to be saleable.

This may be variously from six weeks to four months, according to circumstances, and the care expended upon them.

## Fish Culture in the Parlor

Though the instructions detailed above were given with reference to cultivation in the open air, yet by following them the same thing can be done in the parlor, but on a small scale, and with limited resources. The writer has often delighted his customers by furnishing them with sprigs of water plants, upon which eggs were adherent. These they hatched in a glass tumbler, placed upon the window sill, gradually increasing the size of the vessel, and consequently the quantity of water as the young grew up, until they were finally ready for permanent residence in the aquarium. Such specimens are generally looked upon with great pride by their owners because they are home-bred and grew up in the midst of the household, where all could watch them and learn a bit of natural history without an effort. The spawn for this purpose can frequently be obtained from any well managed aquarium, if the habits of the goldfish are understood.

# 15 Care of the Ponds

It being the intention to raise goldfish for aquarium purposes, this final end must never be lost sight of, for with that end in view we direct all our efforts to make the net result in every way satisfactory. To that effect the water in which they are grown should have no current, neither must there be a continuous supply of fresh water from the outside. The natural habitation of the goldfish is standing water, and if they are cultivated in like conditions, they will the better be fitted for the life of confinement in the aquarium.

It is only necessary to add water to the ponds now and then, just as the fluctuations of the season may dictate, and only in quantities sufficient to preserve a uniform height in the ponds. When additional water is required, it should be turned on from the supply drain, in the day time only, as one can then watch it better and keep out any extraneous matter that may happen to be in the drain.

Under no circumstances should it be allowed to run in during the night, nor in the absence of a reliable person who could turn it off in time, in case a storm should come up. Any possible damage to the ponds can be prevented by keeping out the accumulated water during a heavy fall of rain.

The supply drain should always be kept free from obstructions of any kind, and especially when a storm is approaching, it ought to be examined to see that it will quickly carry off the rainwater without interruption.

The grass and weeds that grow along the edges of the dams are to be kept closely cut, for, if permitted to remain, they not only detract from the appearance of the establishment, but they afford excellent shelter for the numerous enemies that constantly threaten the fish. The ponds themselves require a daily examination, and anything found in them that is not wanted can be removed. Bits of cut glass or leaves from neighboring trees do not harm anything particularly, but they are liable to clog the outlet and cause the water (should a storm of rain arise) to rise to an undue height.

Besides these, there will often be found insects, larvae, etc., which can be removed with a dip net, while other and larger enemies may appear that will require the services of a trap or the exterminating influence of fire-arms.

When the fish have been taken out of the ponds in the fall (which is done by draining off the water, to be described elsewhere), the soft mud is removed from them with a hoe. This mud, when frozen thoroughly, will make a first-class compost for flower beds in the following spring. The wire guards from the inlets and outlets, together with the adjustable sections from the drain pipe, are taken into the house for safe keeping during the winter, the ponds being left in a dry state until again needed the succeeding spring. The frosts of winter will kill any remaining vermin, purify and fertilize the soil of the bottom while the absence of water will offer no inducement to muskrats, whose advent would greatly damage the dams.

Any projected changes, alterations, improvements, or the construction of new ponds should be completed in the fall, so that everything will be in readiness for the spring; then a late season that crowds spring-work in the fields and garden will be of little consequence, as the culturist is prepared to take advantage of the first coming of warmer weather.

## 16 Fishing the Ponds

For the purpose of taking the fish from the ponds, the water in the latter is drained off. But before this is done two more large-sized clean tin vessels are to be provided—tin buckets or wash-boilers will answer very well. Also two dip-nets must be obtained, the one with a handle about seven feet in length, the other a smaller hand-net. The storage ponds are then prepared for their final purpose, and when this is all correctly done, the movable sections at the outlet of the pond containing the fish are removed one at a time. When the water is sufficiently low, the fish are carefully taken out and at once put into the tin vessels, which have previously been partly filled with clear water, assorting the fish according to size, color, etc., at the same time. When this is completed, the fish are put into their respective quarters in the storage pond with as little delay as possible. It is also at this time that the breeder makes his selection of those fish he wishes to breed from; these ought at once to be put into the winter pond.

During this fishing process it is impossible to avoid making the water muddy; so, to prevent the weakening of the fish, the supply drain is opened, allowing a constant stream of fresh water to flow through the pond.

Again the writer admonishes caution; do the work gently and neatly, as every broken fin or lost scale reduces the value of the fish so injured.

## Domesticating the Fish

When in their respective ponds, it is not necessary to feed the goldfish, as nature provides them with all the food required for their proper growth and nourishment; but when removed to the storage ponds, additional food in small quantities may be given to them. This may consist of stale (but not moldy) white bread, dried in an oven or the open air, and crushed to resemble fine hominy or corn meal. Either of these, or both, in small quantities is strewn on the water; the fish being unaccustomed to it will eat but little at first, neither will they snap at it immediately. Gradually, however, they take kindly to it, and the quantity may be increased, keeping pace with the appetite they evince for it, giving it to them at a regular hour each day.

This feeding is not intended for the purpose of making them grow, but rather to prepare them for the change of diet that will ensue when transferred to their future homes.

When the fish take such food and thrive, upon it, they may be considered domesticated, and can be disposed of as pets, they then being in condition to take readily to the more confined life in an aquarium.

When performing work of any kind on or near the ponds, or other receptacle where fish are kept, or in feeding or handling them, a patient and gentle manner is advisable, as it tends to tame the fish by giving them confidence. On the other hand, if they are frequently scared, they become of a wild, restless nature, and will dart away on the slightest provocation.

# 17 Care of Fish in Stores

When fish are kept in tanks as merchandise by dealers, the location of such receptacle should be well lighted, airy, and not permitted to freeze. As was stated in another chapter, it is the oxygen contained in the water that the fish breathes, and it is therefore necessary to consider what means can be employed to keep up a continuous supply of it. One way, the most in use, although the worst for the retail customers, is to keep a stream of water constantly flowing through the tank. Now, this is wrong. Goldfish are intended to live in standing water, and should not be made accustomed to the contrary, as the reversing again of the character of the water often proves fatal to them.

The proper way to keep fish is in pure standing water, to which the necessary oxygen is supplied by the action of aquatic plants; these every dealer in fancy fish is compelled to keep on hand, if he understands the principles upon which the aquarium is managed, and if he wants to make the handling of fish a financial success.

Aquatic plants, when in a healthy condition, exposed to the light, consume the carbon in the carbonic acid gas which is produced by the fish as refuse matter, and give off the oxygen, which in turn is appropriated by the fish. This answers the question as to how many fish can be kept in a certain tank, for it is easy to understand that a locality favorable for the growth of plants will produce the greatest amount of oxygen in the water. Large fish consume more oxygen than small ones, so the proportion of fish to the tank must not be greater than the supply of oxygen the tank can produce, Besides all this, a light location is more beneficial to the color o the fish, and also affords a better control of the contents. Such a store-tank may consist of a large aquarium with glass sides, or it may be a wooden trough thirty (30) inches in width, ten (10) or twelve (12) inches in depth, and of any convenient length. The frame to form the sides and ends is made of one and one-half (I inch stuff, the bottom being formed of flooring boards, as they are fitted with tongue and groove. To make a tight job, pieces of flag leaves, such as are used by coopers, are laid upon the edge of the frame, and the strips of flooring board nailed down securely, one at a time. The groove in each piece is thickly painted with pure white lead ground in oil, the tongue of the next then being tightly fitted into it, and so on, piece after piece, until the bottom has been completed. The best way, probably, is to nail the flooring crosswise upon the frame, as that makes the trough very strong and capable of carrying a considerable weight of water, the smooth side of the boards is of course turned towards the inside of the tank.

Such a tank, however, when in operation should not contain a greater depth than six (6) inches of water, rather less probably, both for the convenience of catching the fish and the better admission of light. The trough must be kept clean, every now and then removing all the contents and thoroughly sponging the interior.

This may appear to contradict the author's method of managing an aquarium; it may be said in explanation that dealers' tanks are, as a rule, overstocked, and therefore require a somewhat different treatment; they are, in comparison with a regular aquarium, the same as a hotel is, compared with a private residence.

While in the hands of the dealer the fish should receive a limited but regular supply of food, and should disease make its appearance, the sick fish are at once taken out and put by themselves.

# 18 The Care of Goldfish in Aquaria

An aquarium may be made of any water-tight vessel, the material of which will not alter the qualities of the water by impregnating it with anything that would stain it or give it an offensive odor, either of which would be detrimental to the health of the fish.

The shape of such a vessel is usually determined by the requirements of the duty it is expected to perform, and should not, therefore, be wider at the top than at the bottom, for then the fish would leap out; neither should it be too narrow at the upper part, for this would exclude the atmospheric air by a reduction of the water surface. The old-fashioned fish-globes are about the worst vessel that can be selected for the keeping of goldfish as pets; they will do well enough for a temporary display of the fish, but for permanent use they may be compared with the Black-hole of Calcutta. Besides they are extremely dangerous, in regard to their round shape, as there are several cases on record where fish-globes, hung near a window, set fire to the lace curtains or the carpet, the globe acting like a condensing lens.

When the vessel in which to keep the pets has been selected, it is thoroughly cleansed with water only, the bottom is then covered with welt-washed river, or white sea-sand, to a depth sufficient to allow the planting of one or more varieties of aquatic plants. The number of these from which to make a choice selection is large, but in these pages only the leading and more desirable ones need be mentioned, the list being arranged according to their relative value as oxygenators. For use during the winter season, we select all or part of the following varieties:

Myriophyllum spicatum.
Sagittaria natans.
Ludwigia Floridae.
Ceratophyllum rosaea.
Cabomba viridis.

Either one of these, grown as a single specimen, if of sufficient size, will do the work of aerating the water during the colder season.

If the tank is of a sufficiently large size, specimens of all of them may be introduced, and as all bear differently shaped and colored foliage, they will greatly enhance the attractive appearance of the aquarium. For the summer season we may make our choice from the following:

Ceratophyllum demersum.
Ceratophyllum roseae.
Anacharis canadensis.
Vallesneria spiralis.
Sagittaria natans.
Potomogeton crispus.
Potomogeton pussilus.
Cabomba rosaefolia.

*Aquarium*

In both seasons we can add to the aquatics named, various kinds of marsh plants and those aquatics that float upon the surface, all of which add to the ornamental beauty of the collection. It may be mentioned that the plants put in a vessel containing fish are not placed there for the fish to eat, as many suppose. Some fish, however, have the habit of biting and tearing the plants from a spirit of mischief, very much like the restless horse that gnaws its crib.

This destruction of the plants can be obviated by placing in the aquarium such varieties as the fish objects to, for it is a fact that they manifest a liking for one kind and a repugnance to another.

The roots of the plants are imbedded in the sand, and bits of rock or pebbles placed around them to retain the plant in place, so that it will not be shifted about in the water. The vessel is then filled with "pure drinking water" to within a couple of inches of the top (See illustration); the water, if not fit to drink, can not be fit to be put in the aquarium, so that it is very essential that it be of prime quality.

When the water is in, one or two frog-tadpoles and a couple or more of pond snails are put in for the purpose of consuming any decaying vegetable matter that may appear, and to keep down as much as possible the growth of confervae; the number of tadpoles and snails must be determined by the size of the vessel and the rapidity of growth of the alga, etc.

When all these preparations have been made, the aquarium is ready to receive the fish.

Before they are put into it, however, the temperature of the water in the vessel in which the fish are brought must be equal with that in the aquarium. This is easily accomplished, and does not subject the fish to any risks that would either make them sick or be fatal to their lives.

When the temperature has been equalized, the fish are gently introduced to their future home, taking care that they are not plunged in so roughly that they become frightened. This performance may, under some circumstances, consume an hour's time, when, for instance, the fish have been carried a distance during severely cold weather, as then the changing of them from the transporting vessel into the aquarium must be made with great care.

The shape of the aquarium and the location in which it is placed determine the number of fish that can comfortably live in it. Should the location be bad as regards light, the amount of oxygen generated in the tank will be less, while if the situation is highly favorable in every respect, the evolution of the life-giving gas will reach its maximum degree.

The quantity of water required for a given number of fish is regulated by their size and the nature of the treatment they have received before they came into our possession. If, for instance, they were raised or had been kept in running water, or were newly caught in a large pond, they will naturally require a much larger quantity of water than if they had already been accustomed to a life of captivity.

As a rule, and It is a reliable one, each fish of three or four inches in length should be supplied with a gallon of water in which to live. Of course this norm may be disregarded for a limited period with impunity, but for the continuous welfare of the fish it must have its proper share of water.

The location of the aquarium should be such that the plants in it will be stimulated to their full capacity of growth, as this is the prime factor upon which depends the maintenance of the aquarium in a proper condition.

The subject of food is one that is but little understood by the majority of people, and is also a matter of no little importance. Most persons, in their anxiety to supply their pets, greatly overdo the thing; the waste material accumulating in the water, remains until it putrefies, thus polluting the water and rendering it detrimental to the health of the fish.

The feeding time should be but once a day, and that at a regular hour, the food to consist of flies or prepared fish-food, such as is obtained from the dealers. The quantity administered ought not to exceed what they will immediately consume. Once a week they may be given finely chopped fishing-worms, or raw beef scraped from the piece, but only in such quantities as to allow each fish a small mouthful. If any remnants of the last meal are found in the water, they should be removed at once, and the feeding entirely suspended for one or two days following. In cold weather the goldfish has but little or no appetite, while on the other hand, they eat voraciously in the summer. The best temperature for the water is somewhere between 60° F. and 90° F., though the fish can stand it as low as 32° F. and as high as no° F. without injury, if the change is not suddenly made, and a corresponding supply of oxygen present. Cold water retains the most oxygen, and also has the power of absorbing more of it from the atmosphere than warm water does. The warm water, however, is most favorable to the growth of plants, so that the quantity they furnish fully makes up the difference. It is poor philosophy to put ice into the aquarium to reduce its temperature, and it is equally foolish to wrap an aquarium in a bag and allow ice-water to drip upon it, as this is hardly the thing to do if the person cares anything for the parlor carpet.

When the fish come to the surface of the water to breathe, it is a sure indication that the oxygen has become exhausted. A fresh supply is easily introduced by simply stirring the water with the hand or dipping it up with a cup and pouring it back again, it is much better to do this than to put in a lump of ice.

The frequency with which an aquarium is to be cleansed depends altogether upon circumstances and the individual taste of the owner. An aquarium kept in the parlor of a suburban residence, and in a locality surrounded by flourishing shrubbery, needs re-arranging twice a year only, viz: in the spring and fall. On the other hand, an aquarium located in a smoky city and kept in a close, badly ventilated apartment, must be emptied frequently and a new supply of fresh water put into it.

Some people object very strenuously to the formation of algae upon the glass sides, and on that account clean the vessel very diligently, removing every particle they can find. Others again change the water because it does *not* form, their taste preferring the fish in all the natural surroundings of a rural locality.

The former carefully wipes each pebble till it shines, and "would polish the fish too if it were possible, while the latter will walk for miles to some creek in order to procure some moss-covered rocks; so divergent are tastes in this matter.

One must be able to exercise his own judgment as to the best time and when, for the changing of the water in the aquarium, as it may sometimes be better to leave it undisturbed for some length of time, and at others to change it several times.

The best means to clean the glass sides from the adhering algae, when an aquarium is emptied, is by the use of a rough sponge or rag dipped in whiting; this will remove every speck without scratching the glass.

The best side of a room for the aquarium is that having a window, near which it is to be placed, as the light can be increased or reduced by opening and closing the shutters.

In the winter this position is the best, for the constant ventilation that goes on in the immediate vicinity of the window protects thee water from the injurious effects of coal and tobacco smoke, and the poisonous fumes from the gas-burner; in the summer, in close hot weather, before a thunder storm, it can be easily and effectually aerated.

There remains yet one point: "Should the sun shine upon the aquarium?" This too is a matter of taste, though we would recommend a middle course, that is, let the fish have the sun part of the time. In winter allow them to enjoy the full light of the sun, but towards spring and during the summer shelter the tank from the direct rays.

If the goldfish are kept in water that contains no aquatic plants, it will have to be changed frequently as the oxygen in it can not, in most cases, be replaced from the atmosphere as rapidly as it is consumed by the fish.

# 19 Transportation of Fish

The vessel best adapted for the shipment of live fish, to any reasonable distance in this country, is a tin can, clad with wood.

The shipping can should be perfectly smooth upon the inside, so that the fish will be subjected to the least percentage of injury while *en route*. At the upper end the can should taper off, forming a kind of neck or shoulder, similar to that seen upon the common coal-oil can. This makes it easy for the contents to slide out when the vessel is to be emptied. The opening at the top is five (5) inches in diameter, and closed with a perforated lid that is fitted in like the top to a milk can. The perforations consist of half a dozen one-half (%) inch holes, punched through from the under side of the lid, thus leaving the sharp rim of the holes on the outside where they can do no injury to the fish within the vessel. (See illustration.)

Those cans very extensively used in the coal oil trade, and usually designated "wooden jacket cans, "are about the very best thing that could be invented for our purpose.

*Shipping can*

As the Express Companies demand that tin vessels be protected in some manner or other with wood, we find in these vessels the fulfillment of all requirements of that nature, and at a slight increase in weight. Besides, these cans are readily obtained in the large cities, (and smaller ones, too, in all probability), the only necessity being the enlargement of the opening to make it a complete shipping can for fish.

When it is desirable to make a shipment of live fish, it is necessary to take into consideration their size, the length of time they will be upon the road, and the season of the year in which they are transported, all with reference to the all-important supply of oxygen, without which, of course, the fish cannot live.

The shipping can is filled with pure water, to four-fifths (4–5) of its capacity only, thus providing ample space for the water to splash about during the journey, as it is by this constant motion of the water in the vessel that it is aerated and made capable of supplying the fish with oxygen.

The hour of shipping ought, if possible, to be so arranged that the journey on the road may be made at night, as it is cooler in the summer, does not expose the fish to the great heat of sunlight and the arrival is made usually sometime in the morning or forenoon when those at the destination are on hand to receive them. The cans must be plainly labeled, stating the nature of their contents, so that they may receive more care in the handling from the express agents, consequently running less risk of damage. It is also advisable to notify the party to whom the fish are sent that the shipment has been made, in order that he may take them from the agent as soon as possible, otherwise, if they are left to remain quiet at the express office or freight depot, the fish will be in great danger of their lives from want of proper care.

If the shipping can is an ordinary small tin bucket, such as are on sale at the tinsmith's, the ventilating holes would better be punched in the center of the lid, the remainder being left unmolested, forming a shoulder against which the water can splash without being spilled. In all cases the lid must be securely fastened with strong twine or wire, so that a jar will not displace it.

The above directions apply more especially to shipments that do not occupy any great length of time in The transit.

If the fish are to be sent great distances over our own country or exported to foreign parts, the safest plan is to put them in a vessel fitted up like a regular aquarium. Japanese goldfish have been sent to Europe with perfect success in the following manner:

A one-gallon candy-jar, (such as are used for hatching spawn) is fitted up in proper style, with sand, water-plants, snails and tadpoles, and filled nearly to the top with pure water. In this may be placed four two inch fish, the top then covered with a perforated tin lid, and the whole set aside for observation for about a week. During this time a tin bucket is obtained of such a size that the entire candy-jar aquarium will nicely fit into it, the top of the jar being neither higher nor lower than the upper edge of the bucket. This tin bucket, or sheath, if you will, serves as a perfect guard against breakage, and *should any* accident occur to the jar, the bucket is on hand to act as a substitute. Furthermore the tin is provided with a convenient handle to carry it by, and for greater security the jar can he retained in its place with a heavy wire bar across the top, so adjusted that it can be removed at will.

On board the ocean steamer, the buckets are suspended by the handles, the water is not changed, neither are the fish fed anything.

As will be seen, the uncovered shoulder of the jar, will admit plenty of light, so that the plants can act on the water and keep it fresh, neither can the water in the jar be lost by splashing out, as when this does happen, it merely falls into the bucket, from whence it can be returned; the perforated lid admits the air, but at the same time prevents the accidental escape of the fish.

This description of a trans-Atlantic shipping can is not at all expensive, and is further recommended by its reliability. Make it a rule, however, to ship only such fish as are in perfect health and fully domesticated.

While on the road the water in the can should be changed only in exceptional cases, and then with great care. If the fish become weak it is a sign that they are not in good condition, and that a mistake has been made in preparing them for travel, and the simple changing of the water then will not prevent their dying. The rule is, do not crowd the shipping cans.

When, upon the arrival of a lot of fish, there happen to be any dead ones in the vessel, and the balance weak or in a dying condition, or look slimy and pale, with bloody streaks on the fins or around the scales, it is a sure sign that suffocation has been the cause of the death of some, and will speedily cause that of the others. The living ones should at once be placed in a large vessel in the open air, filled with fresh water, to which a good handful of common table-salt is added. A clean wash-tub answers the purpose nicely, and besides, has the merit of usually being close at hand.

This treatment, if resorted to immediately, will, in most cases, restore the fish to good health.

# Part IV

# ENEMIES AND DISEASES OF THE GOLDFISH. REQUISITES, TOOLS, ETC.

# 20 The Enemies of the Goldfish

*"Knowledge is power."*

It has been stated elsewhere in these pages that the raising of goldfish consists largely in the protection of them against their enemies, and to make the defense most effective it is essential that we know something about the transgressors, for by being acquainted with them and their habits, we can more intelligently combat them. For this end the author has described them in the following lines, adding to those that are less familiar to the general reader, an outline of their natural history and viewing them in the successive stages, during which they endanger the life of the fish.

Before beginning the description of insects, however, it is proper to say that the spawn of the goldfish immediately after its deposition, is sought for by other fish and devoured, the spawners themselves also engaging in this nefarious practice.

Those eggs that have escaped the notice of the fish are consumed by various smaller enemies.

## The Common Pond Snail

(*Lymnea fragilis.*)
Foremost among these enemies to fish is the common pond snail, which devours them.

## The Water Asell

(*Asellus aquaticus.*)
This little creature, of which the accompanying illustration gives a good idea, is a crustacean, not more than one-half of an inch in length. It crawls about upon the bottom of the ponds and over the water plants, searching for food, part of which consists of fish eggs, to which they are very destructive, devouring them wherever found.

Another voracious enemy of the crustacean tribe is

## The Water Flea

(*Gamarus pulex.*)
The water flea, or Flea Crab, Buck Crab, etc., is a lively little creature closely related to the shrimp found in the ocean. It furrows through the water in any direction lying on its side, because its back is naturally bent. (See illustration.) In all stages of its growth it feeds upon the fish eggs, but in turn, furnishes an excellent food for young fish.

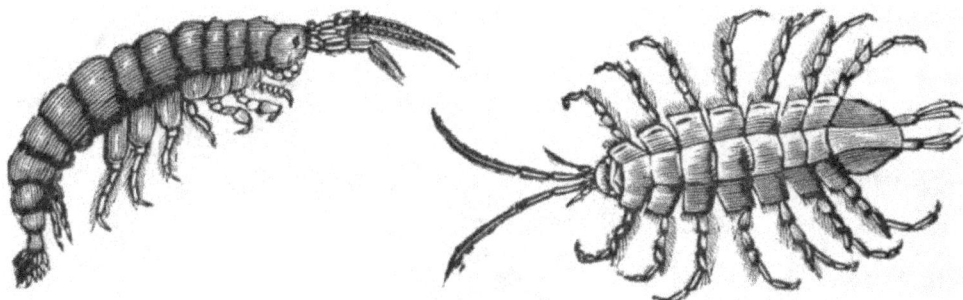

*The water flea (left) and water asell (right), enlarged*

## The Boat Fly

(*Notonecta glauca.*)

The boat fly, or "Shoemaker, "as it is commonly called, is a most voracious insect. The body is long, contracted posteriorly, convex above and flat below, having hair at the sides and extremities, which, when spread out, supports the insect upon the water. The head is large and presents a large eye upon each side, giving the possessor the power of vision in all directions. The color of the body is a greenish grey, the wings are white, of the legs, the four nearest the head are short, but the third pair are very long, different in shape from the others, very much resembling boat oars. When in the water, the insect swims upon its back, using the hind legs as oars for propulsion, while the front ones are instrumental in seizing its prey. Young fish, tad-poles, and other insects, all contribute to supply it with food, to the former, especially, it is a very dangerous enemy. The instrument or weapon with which the insect makes the attack upon the victim is a strong, conical beak.

It is believed that when making the attack, the boat-fly injects poison into the wound it makes, as seems to be proven by the fact that when once attacked, though subsequently escaping, the

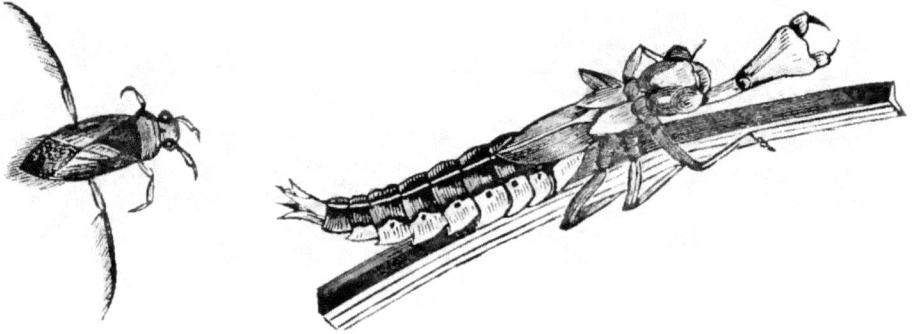

*Boat fly (left) and larva of dragon fly (right)*

victim always dies in a short time. When upon land, this fly crawls along, in an upright position, dragging its oars behind it. In the evening, and at night, it likes to leave the water and make excursions to other ponds or creeks; from this habit the culturist may take warning. Its eggs are deposited against the stems of aquatic plants in the early spring, and again in mid-summer, so that one season produces two crops of them.

The young make their appearance soon after, immediately following the example of the parents by swimming upon the back and eating almost anything they happen to meet. The accompanying illustration shows the insect as seen from below when in the water.

There are two or more varieties of this fly that differ in coloring, and of smaller size than the one described, though all are extremely destructive to the young fish—the one just delineated, more especially.

## The Yellow-Banded Water Beetle

(*Dytiscus marginalis.*)

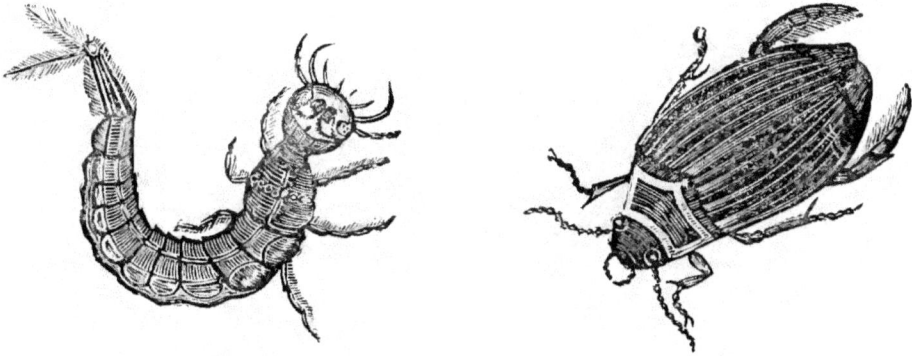

*Yellow-banded water beetle (right) and its larva*

This rather pretty beetle lives entirely below the surface-of the water, never leaving it, except during the night when the air is damp or in rainy weather, and then for the purpose of making excursions to other localities. The body is of a greenish black color, encircled with a brownish yellow band—this feature giving it its name, When taken from the water it exudes a milky fluid of a most offensive and disgusting odor. The hind legs are shaped very much like those of the boat-fly, and serve the same purpose. This beetle is very courageous, attacking fish of my size, as large ones have been caught, into whose flesh the beetle had eaten large holes, the beetle itself found in the hole hard at work eating up the fish. The larva, which is produced twice within the same season, lives and grows upon tadpoles and young fish.

When of sufficient size, and the proper time has arrived, it changes into a pupa, which in turn, becomes the perfect beetle. (See illustration.)

## The Black Water Beetle

(*Hydrophyllus piceus.*)

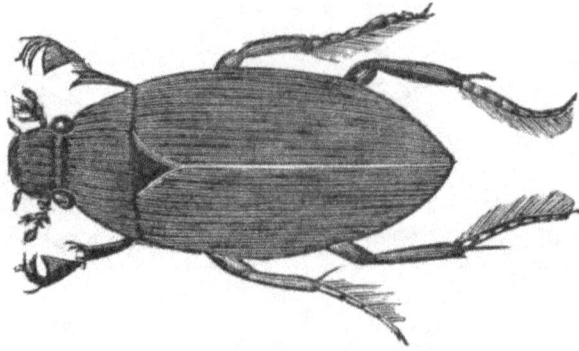

*The black water beetle*

As the name indicates, this beetle is black, shining with a rich, purple lustre. (See illustration.) It is of larger size than the preceding, and strong in proportion.

The beetle itself is a vegetarian, and as such, is not directly dangerous to the fish, its larva, however, is voracious without limit, destroying all that comes in its way.

The female of this species spins a white cocoon around the posterior portion of its body, with the aid of its hind legs, the cocoon, when completed, being the size of a hazel nut. In this it deposits its eggs, and after closing it carefully, fastens it to a floating leaf, adding to it a little projecting point on the top, which by the way resembles a small mast, retires to the water underneath and mounts guard. After a few days the young grubs make their appearance, at first resembling little whitish worms, but possessing six legs near the yellow head.

It is by the motion of these legs that the grub is propelled through the water, continually on the search for something to eat. When at rest on a water-plant, the head with its fearful apparatus, formed of a strong pincher with two pairs of adjuncts, which can be moved in any direction, is placed in such a deceiving position as to almost always lure an unsuspecting little fish, tad-pole or insect, within its reach.

As the grub gets larger, it turns darker in color, until having attained a size of about four inches in length, it has become nearly black on the back; the under part is then of a creamy white, and the sides have been fringed with hair. In this state its appearance is extremely repulsive, being about as ugly as anything can be imagined. The earliest and best time to destroy them is when the cocoon has been finished, and the female is standing guard in the water beneath, both can then be captured and obliterated, in this way great damage is prevented before there has been an opportunity for development; very much on the principle of the old proverb: "A stitch in time saves nine."

The grubs breathe through the posterior part of the body, and have to come to the surface occasionally for that purpose, at which time they are easily caught with a dip-net.

In general appearance, the color excepted, the grub of the black water-beetle resembles that of the preceding.

The beetle, moreover, is very prolific, spinning several cocoons at two different periods, namely, in the spring and high summer.

Other varieties of this insect exist, the one under discussion being the most dangerous to the fish. So far as the others are concerned, it is sufficient to remember the injunction, allow nothing alive to remain in the company of the fish when newly hatched.


**The Dragon Flies**

The dragon flies (commonly known as snake-feeders) may be divided into three classes, all very destructive enemies of the fish.

1. The *Libellula* possesses a short, flat body, about two inches in length. (See illustration).

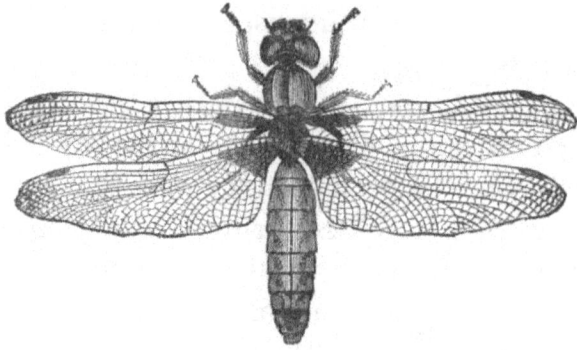

*The dragon fly*

2. The *Aeshma* is longer than the above, its slender, round body sometimes measuring six inches in length.

3. The *Agrion* is not large, the body small and slender, varying in length from 1 1/2 to 234 inches.

The wings of the first two named, are, when the insect is at rest, always expanded horizontally, while those of the latter are folded together, pointing backward.

The hind part of the body in all of them is long, slender, and composed of ten rings. On the forepart of the body, they have three pairs of legs, and two pairs of transparent, webbed wings, the latter in some species glitter like gold, in others they are dotted with spots of different color; in the Agrion species they are of the same color as the body. The coloring of the bodies of all, especially the aeshma, is very brilliant, being of a bright green, blue or scarlet, and sometimes mottled and spotted with various colors.

The eyes are large and prominent, giving the insects a very large field of vision.

They all fly very rapidly, feed upon insects of every description that they catch flying about, and from this fact they may be made useful to destroy the mosquitoes in bedrooms and elsewhere. Although very voracious, they are perfectly harmless to, man—they can not injure him in any way. The manner of their copulation is somewhat curious. The male fastens the extreme back part of its body to the neck of the female, and thus attached, both fly about for one or two hours, when, over some water, they separate. The female then deposits her small white eggs by immersing the posterior part of the body in the water, attaching them to the submerged surfaces of water-plants; there they remain until hatched.

The larva; or grubs of the dragon-flies live in the water; those of the libellula are short and thick, while those of the other genera are more slender, corresponding with the shape of the adult. The color of these grubs varies from blackish-brown to a brilliant green. They breathe through the posterior part of the body, which apparatus is also used to propel them forwards through the water, making them good swimmers.

They are extremely destructive to young fish and fish-eggs, upon which, together with tadpoles and snails, they manage to make a good living., Instead of hunting their victims, they lay concealed in the mud with the eyes only protruding from the surface. Whenever a victim comes within reach, they produce their concealed pincers by a rapid motion, rarely missing the mark they aim at. (See illustration.)

There are instances on record where one of the larvae of the libellum, which was overlooked in the fish-tank, destroyed two thousand (2,000) young fish in a week's time.

After they have attained their full growth, the grubs leave the Water, climb upon some object projecting from it, when the perfect fly makes its appearance through the back of the grub, rising upon its wings into the air as soon as they are unfolded and dry.

The eggs are also produced twice in a season, the grubs from the last deposit, living in the mud during the winter, and produce in the early spring the first dragon flies of the season.

Their natural enemies are the frog and the water-spider. The latter, small as it is, compared with their own size, is, nevertheless a powerful antagonist, attacking them when in the act of depositing their eggs. The attack is made upon the eye, the largest dragon-fly thus being easily overpowered by its small but intelligent enemy.

How strange it is that just those animals with which man has the least sympathy are among his best friends! Such are the toad and the spider!

# 21 The Enemies of the Goldfish, Continued

## The Cray Or Craw-Fish

(*Astacus fluviatilis.*)

The crayfish is also known as the fresh-water lobster, and should be killed whenever and wherever met, as it is very destructive to the eggs of fishes.

It will also occasionally catch a young fish, and often injure others by snapping at them, tearing away parts of the fins and flesh, thus rendering a beautiful and valuable fish wholly valueless as merchandise.

The main damage done by the craw-fish, however, is the undermining of the dams, which is not only annoying and costs much time and labor for repairs, but makes it possible for the fish in the several ponds to get mixed by passing from one to the other.

**The Water-Frogs**

(*Ranae var.*)

As the heading indicates, there is a recognized distinction between frogs, there being water-frogs, tree and grass frogs, the latter in no ways molesting the fish in the ponds, in reality on the contrary, making themselves very useful to the culturist by destroying harmful insects. The common frog is the one we have to guard against, both itself and all its varieties, whose destructive habits far outweigh the little benefit derived from them. Just as soon as they have completed their gradual metamorphosis, and become perfect frogs, they prey upon anything that has life, including young ducks, turtles, snakes and cray-fish, as well as fish. Besides this, the adult deposits its spawn in the ponds, and when the tad-poles are hatched, they consume a great percentage of the natural food found in the pond, thus depriving the young fish of proper nourishment, in this way being indirectly injurious.

The frogs are furthermore dangerous, through their habit of wandering in the night from one locality to another, during their spawning season, and thus often unconsciously introduce into the ponds, the spawn of minnows, which being adhesive, sticks to the skin of the frog, and is, of course, carried about by it.

The frogs make their appearance early in the spring, and it is at that time that they can be most effectually destroyed in the following manner: water is let into one of the ponds, to the height of several inches, in this pond they will collect at night, in order to deposit their spawn. Next morning, most of the frogs themselves can be caught with a dip-net, and the spawn also removed and exposed to the sun to dry up, by merely placing it upon the ground, where the sun can reach it. If this process is systematically carried out, during their spawning season, not many frogs will trouble the establishment during the season. When it appears that all are captured, the pond is drained off, so that any tad-poles that might have been hatched in it will die by being dried up in the sun.

In destroying the frogs the culturist must be careful not to mistake the common American toad (Buffo americanus) for the frog, as this innocent creature is worthy of our protection. They may very easily be distinguished by their color, which is brownish and yellow, the skin moreover being warty. Their eggs also differ from those of the frog by the manner in which they are joined together; those of the frog are found in one compact mass, a lump, in other words; those of the toad, in strings; the eggs of tree and grass frogs in sheets. If the amateur will bear these distinctions in mind he will have no trouble ridding himself of a pest and preserving a friend.

It will pay to remove the toad-spawn carefully, and put it into a pond where it can hatch unmolested, which is completed in June or July.

Toads in the neighborhood of ponds and gardens are a blessing, and should not, therefore, needlessly be exterminated.

One of the natural enemies of the frog is the water-spider, which attacks the young in the eye and kills it. Although the water-spider has been repeatedly mentioned as a friend, it is not out of place to keep an eye on him, as he also frequently catches young fish.

### Salamanders and Newts

All the varieties of those tailed batrachians frequent the water in the spring, for the purpose of depositing their eggs or young—some being viviparous. In all the stages of their growth, from the tadpole state to the perfect animal, these creatures are destructive to both the fish eggs and the young fish, they should therefore be kept out of the ponds.

The newts spend their entire life in the water. In certain localities they may exist in such vast numbers that it is necessary to
take especial precaution to keep them away.

Neither of these creatures, however, is harmful to man, the
larger varieties living in the river, such as the *water-dog* and the
*hell-bender,* excepted, nor are any of them poisonous.

They may, with perfect safety, be handled with the bare hand.

As the season advances various other enemies, in addition to,
those already mentioned, make their appearance. Most particularly must a sharp lookout be kept for

## Water Snakes

These will be found concealed near the water's edge, or in
the corners of the several ponds. They lie hidden from view, the
head only exposed, all the while playing the tongue in the water.
This they do to allure their prey within reach, the fish mistaking
it for a worm rush to their certain destruction.

The best means of exterminating them is by the use of some
kind of fire-arm. A smooth bore, 22 caliber Flobert gun, loaded
with a cartridge containing shot, and a good marksman at the
proper end of it, generally makes a combination that forever
prevents that snake from exercising his fishing propensities.

If young snakes are about, they may easily be discovered by
placing pieces of board here and there about the ponds; these
boards are lifted up in the morning, often revealing two or three
of the little snakes that had sought shelter there, when thus
found they are easily dispatched.

A good snake trap was accidentally discovered as follows:

A wire coop made of 1/2 inch mesh galvanized iron wire
netting, served as the dwelling of a couple of muskrats, which
the children kept as pets. These having died, the coop was used

to confine live frogs, and kept outside of the establishment in the water, just at the point it leaves the ponds.

The next morning a large water snake was found caught in the meshes of the wire, and dead.

It had evidently tried to get into the coop, and help itself to a frog or two, and was thus caught, the wire preventing the entrance of the entire body, the scales of the reptile at the same time precluding the possibility of retreat. Since then this trap, and smaller ones, have been used with very good results, and being simple they are easily made.

Following the snakes, and at the time the ponds are filled with water, one must be on the watch for fish-eating birds, among which are

## Cranes And Herons

Cranes and herons, which, seeing the surface of the water below whilst flying above, are attracted by the glisten, and immediately descend to reconnoitre. These birds visit the ponds at regular hours, wade in the water and catch with consummate skill all the fish they can get. The fact that their stomachs are sometimes found to contain nothing else than crayfish must not mislead the amateur into the belief that they prey only on these crustaceans, as the following fact abundantly proves that they prefer fish whenever they can be obtained. A heron was seen flying towards the ponds; to secure the ever-ready shotgun from the lodge, sneak within range and fire, did not take more than five minutes' time. Yet within that short interval the bird had captured and devoured three 2 1/2-inch long, brilliantly colored goldfish, which, though already dead when taken out of the bird's stomach, were still perfectly bright, showing conclusively that they had just been swallowed.

If these birds discover that fishing in the ponds pays well, they will become frequent visitors, and, if not killed, soon clean out the establishment. But, as before stated, they come at regular hours, thus affording the one on duty at the ponds an easy chance to be on the watch to kill or trap them.

### The Kingfisher

This bird may likewise be expected to visit the ponds, but it generally advertises its arrival with a lusty kar-r-r-r-ack! that may be heard quite a distance.

It selects projections over the water, such as a branch of a tree, a post, or the outlet pipe of the ponds, from whence it shoots down upon the unsuspecting fish, seldom missing. It also supports itself upon its wings immediately over the water, darting down upon its prey with like success.

These birds, although not so easily shot as the crane or heron, may readily be caught in a trap, if the latter is somewhat concealed and laid on the post or outlet pipe, which the birds mostly frequent.

### Turtles and Alligators

These reptiles are both extremely destructive; the latter, of course, not being found in the Northern States, need not be looked for in that locality. Neither of them, no matter how small, should be permitted to remain in or near the ponds.

### The Muskrat

Not only do these animals destroy the dams of the ponds, but they will also destroy the entire stock of fish, if not stopped in time. Luckily for the fish-culturist, they are easily mastered. One or two muskrat traps of the old-fashioned style (Hawley & Norton's No. 1), used by professional trappers in the Far West, can be procured at almost any hardware store for thirty cents apiece, the chain included. The muskrat holes are looked for and will be found leading into the bank and a little below the surface of the water. At a short distance from such a hole (the length of the chain on the trap), a peg is driven securely into the ground, and the free end of the chain fastened to it. The trap is then set *without* bait, and laid a little to one side immediately into the hole, in such a manner that the animal in going in or out is obliged to tread upon the plate that springs the trap and over one or the other *end,* thus it is always caught by one of its legs.

If the trap is placed at right angles with the hole, so that the animal has to walk over the bows, these latter, in coming together, will throw the rat upwards, and fail to catch it.

It is advisable also to catch the muskrats in the surrounding neighborhood of the ponds, as they make excursions during the night to the ponds in order to fish.

This long list of enemies may be increased by adding the *raccoon,* the *mink,* and water-fowls such as *ducks, geese,* and *swans.*

In stores where fish are kept in tanks they must be watched and protected from *house-rats* and *cats,* both of which will occasionally make a descent upon the tanks if not prevented.

# 22 The Diseases of the Goldfish

The goldfish, when in perfect health, carries the dorsal fin in an erect position, in other words, fully expanded. Its colors are very distinct, the body of the fish glistening as though highly polished. The fins appear very clear, translucent, allowing an examination of their structure, they are also very flexible moving in the water with animation and grace. When closed by the fish, the gill covers fit tightly against the head.

Liveliness is not always an indication of good health, and, on the contrary, sluggishness is no positive evidence that the fish is ill.

But when the brilliant red color fades away into an off-colored pink, or the milky white portions of the body become intermixed with bloody streaks, or the fins of the fish appear to be coated with something unusual, or seem inflamed and stick together, or are carried close to the body, or when the gill covers appear so swollen that they will not fit tightly in their proper place, then the health of the fish has failed, and danger is close at hand.

Most of the diseases of the goldfish are the direct result of ill treatment while kept in captivity, and nearly always originate in the breathing apparatus; the gills, when affected, fail to supply the blood with oxygen. Some of the diseases, to which the fish are subject, originate from improper methods of feeding, and always manifest themselves in disturbances of the stomach, and other digestive organs.

Again there are diseases that make their appearance periodically, the origin of which is involved in as much obscurity as that called "pink-eye, "which attacks horses, and of which all have heard more or less.

## Asphyxia

This affection is the one most commonly met with. The fish become weak, the colors fade away rapidly, the appetite is lost, and the fish finally die if the disease is permitted to run its course without hindrance.

The cause of the disease may be looked for in the interrupted functions of the gills. These organs become inflamed by the irritating and poisonous gases that may exist in the water, or by the sudden changes of temperature in the same.

If the disease has not already advanced too far, the ailing individual or individuals, should be taken from the collection, placed in a vessel containing a sufficient quantity of water, and in which a number of flourishing aquatic plants are growing.

An even teaspoonful of common salt is then dissolved in the water, the whole then put in a light, well ventilated place, and kept at a temperature between 70° and 80° F. During the first few days no food is necessary, and should not be given, after which the feeding may be re-commenced, beginning with very small quantities, administered at a regular hour each day. As the fish brighten up, and approach convalescence, the quantity may gradually be brought up to the usual amount.

### Tuberculosis or Consumption

The first sign of the presence of this disease shows itself in the indifference manifested by the affected individual. They are seen swimming about in a careless, purposeless way, now and then stopping to make the vain attempt to remove something from their gills that annoys them. They are apparently coughing. Their appetite decreases. It is evident that the gills are out of order, they thus failing to take up oxygen for the blood.

As the disease progresses the fish becomes lean, as seen back of the head, on the back, and the sinking in of the abdomen, causing the head to appear too large and out of proportion. The gills become agglutinated which results in the destruction of their structure by decay. Having arrived at this stage the fish is too weak to balance itself and swims head downward, finally standing on it, because it is the heaviest part of its body, and dies in that position.

The duration of this disease varies in different individuals and seasons, it being of shorter duration during cold weather.

The origin of this disease may be traced back to *unnatural treatment* while in captivity, as fish in native waters never get it, and in well managed aquaria they very seldom have it, while in those badly managed they frequently die of this disease. The cause of the disease arises from invisible organisms called *tubercular baccilla*, these being inhaled by a fish whose breathing apparatus is in the least out of order, infest these organs and destroy them.

The disease proves fatal in all cases, or has invariably done so in the writer's experience, who has so far failed to discover an effective remedy. The disease may be avoided by keeping the aquarium in perfect condition.

## Slime or Itch

This disease generally appears during the colder season of the year, seldom manifesting itself when the weather is warm or hot. The body of the fish becomes coated with a layer of some whitish substance, the deposit beginning on the back near the head. This white, slimy substance, when examined under a powerful microscope, reveals a number of parasites darting about hither and thither across the field of the instrument. These minute organic bodies, technically termed *bacteria*, resemble a wood-tick in general shape. They appear to eat into the skin of the fish, destroying that structure, and in consequence interfere very much with the function the skin performs in throwing out poisonous substances that form in the tissues inside. The beautiful colors of the fish disappear from the tainted parts, they becoming quite black. The result of the disease is the death of the fish.

When afflicted with the trouble, the fish can be observed rubbing itself against the plants, the rocks, or in the sand upon the bottom of the aquarium.

The cause of the disease may be found in the usual overfeeding, in which case the remnants of food remain in the water until decomposition sets in. The temperature not being favorable for the production of water-purifying insects, the *bacteria* make their appearance greatly to the detriment of the fish.

The name "slime "has heretofore been applied to this description of disease, but without any apparent reference to the cause producing the trouble. In the author's opinion, it is the presence of the parasites that makes the whole difficulty. They annoy the fish by their presence, their attacks upon its skin setting up an increased flow of blood to the part upon which the thick coat of slime is found. Manifestly the best method of treating the disease is the removal, or rather the prevention, of the cause, for if the *bacteria* are not allowed to develop, they of course can do no harm.

Tadpoles and snails should be put into the aquarium to consume any remaining particles of food, and the feeding itself more carefully attended to. Place the aquarium in a warm and light location, adding to the water a pinch of table-salt when filling the vessel.

## Dropsy

Dropsy, as every one knows, is a swelling up of the body, caused by the presence of watery fluid in the tissues, so it is with fish when affected in this way.

It generally begins near the tail, but sometimes about the middle of the body and progresses forward. When it first makes its appearance, a few scales in a circle around the body lose their firm attachment, at this stage, if the affected specimens are immediately removed and placed into water brought from some other locality than that in which they had been when taken ill, they will recover in a short time.

The disease having started, will, if not immediately attended to, spread over the entire body until it becomes almost spherical, so great is the distention of the skin. The scales become erect, giving the fish the appearance of a "ruffed grouse," the eyes at the same time being greatly protruded from their sockets.

During all this while the fish shows a good appetite, and continues to do so until the end, which soon follows.

No cause, as yet, has been found producing the disease; it appears upon fish in Europe, as well as in this country, and also upon fish kept in open air ponds, as well as those inhabiting the aquarium, and in any season of the year, and at any age of the goldfish.

The disease may run a course of four months, at the end of which time it results in the death of the fish. It also seems to be intermittent in character, disappearing for several weeks, and returning again upon the same individual, but always in such cases with fatal effect.

There seems to be no other treatment than making the fish as comfortable as possible, taking that chance for recovery.

## Erysipelas

This disease is indicated by what appears to be a nervous restlessness of the fish. They are seen swimming with very quick motions, darting here and there with great rapidity, and with no other apparent reason than a desire to flee from their torment, for it seems that they suffer from muscular pains. After this extreme activity which covers a period of several days, the fishes (for they all become affected at the same time) huddle together on the bottom of the tank, now and then resuming their mad capers.

The external appearance in this case is characterized by a closed dorsal fin, bloody streaks upon all of the fins, which, moreover, instead of being nicely rounded upon their extremities, as in health, become agglutinated and appear like the spikes upon a catfish. The tissue between the spines decays, the latter looking like the disarranged bristles on a brush; this is the beginning of the end.

The appetite continues in good condition, the fish, nevertheless, become lean and weaker each succeeding day until death takes place.

The cause of the disease, also, can be traced to improper methods of feeding, the stomach in consequence becoming overtaxed and the entire system disarranged.

When it does appear, all the fishes are attacked at the same time; the aquarium then should be placed where it will be exposed to the sunlight, the temperature of the locality being kept at about 70° F., and no food administered for about a month. Snails and tadpoles should not be omitted when stocking the hospital tank, as they are excellent scavengers, and by their presence will prevent a complication of diseases.

The diseases described above constitute the main ones we have to expect, and, with the exception of dropsy and tuberculosis, are easily managed, if the treatment is carried out properly.

## Accidents

According to the old saying, "accidents happen in the best regulated families," so will they happen to goldfish. In most cases, nature, if let alone, will repair damages with surprising skill, though a little assistance often helps to secure a desirable result. Scales that have been knocked off will be replaced, just as a finger-nail is when bruised.

Injured fins grow again, but the form afterwards does not always assume perfection.

If an eye has been torn out, it will not necessarily kill the fish, as in most cases it heals kindly, and indeed might (for appearance' sake) be replaced by an artificial substitute, such as are in use by the taxidermist.

When we wish to assist nature to heal a wound, we must bear in mind that a warm temperature is most favorable for that purpose, and is also not favorable for the growth of fungi, which would certainly collect on the wound and reduce the chances of complete recovery. As an additional guard against the formation of fungus, table-salt in quantities mentioned above in treating asphyxia, is good, as also is a solution of carbolic acid, five drops to the gallon of water.

We conclude this description of the various diseases with the homely phrase, "an ounce of prevention is worth a pound of cure," so bearing that in mind, one will see to it that his aquaria and ponds do not suffer from neglect, as that always tells upon the fish.

## Requisites and Tools

Not many of these are necessary for the raising of goldfish. In addition to the ordinary gardening tools used in constructing and repairing the ponds, four sizes of *dip-nets* should be kept near at hand.

1. A large one consisting of a heavy iron wire ring, about the thickness of a lead-pencil, and measuring about fifteen inches in diameter. This ring is securely fastened to a hickory pole seven feet in length, and covered with heavy mosquito netting, the bag to be about twelve inches deep.

2. A medium-sized dip-net of oval shape, measuring six inches by ten inches through the center. This is best made of No. 10 brass wire, fastened to a handle four feet in length, and covered with finer mosquito netting, forming a shallow bag similar to the bowl of a spoon. This net is used to remove insects and small fish from the water.

3. A hand-net of the same size and material as No. 2, with this difference, that the handle may be made of the same wire that forms the frame.

4. A small dip-net, also made of brass wire, No. 16 or 17, in this case of sufficiently small size to use in removing fish or insects from the hatching-jars.

Several one-gallon candy jars for hatching the eggs; several tin buckets of different sizes, for carrying and removing fish; a yard or two of mosquito netting; some wire netting of the same mesh as that used in the guards on the outlets; a thermometer, a couple of muskrat traps, if needed, and a gun to dispose of snakes, birds, etc., complete the outfit that is necessary for the proper performance of the work in hand.

# 23 Profitableness of Goldfish Ponds

The question will naturally arise in the mind of the reader, should he have any desire to engage in the culture of the goldfish, "Will it pay?" "Are the profits accruing sufficient remuneration for the time and labor expended, to say nothing of the capital invested?" To these questions the best answer is the book upon which the author has expended so much time, for he is certain that the proper cultivation of the goldfish will pay, though there are some considerations, the absence or presence of which somewhat determine the result. If looked at from a business point of view solely, it wholly depends upon the local demand for them, though more especially upon the class of customers one is expected to supply. In some places the finer qualities and varieties pay best, as the demand for them comes from a source that is both discriminating and critical, at the same time willing to pay for the very finest that can be had. Such a market is usually found in the larger cities where the wealthy classes generally reside, and it is from among them that the culturist may expect to find a ready and remunerative run of custom. In other places the demand is for quantity not so much care being taken whether the varieties are the best or not; for instance, a gentleman wishes to stock a lake upon his premises or in his garden with ornamental fish, he does not care especially to invest his money in a few choice varieties, but would rather have a greater number of a less desirable kind in order that the lake or pond may contain them in plenty, so that they can always be seen, no matter from what point.

The percentage of young grown to perfection from a given number of eggs depends altogether upon the attention paid to them, the locality in which they are raised, the season in which they are bred, and the variety to which they belong. The number of saleable fish may thus be but ten, or it may be eighty-five, realized out of one hundred eggs. The rules governing the productiveness of ponds are much the same as those that govern the crops raised from the soil.

There is this difference, however, the raising of goldfish is confined almost exclusively to the nicest season of the year, and makes it a highly interesting out-door recreation, and one that can be begun on the most humble scale at trifling expense, besides being an occupation that any lady or gentleman can indulge in with perfect propriety.